B

# DNA IS NOT DESTINY

# DNA IS NOT DESTINY

The Remarkable,
Completely Misunderstood
Relationship between
You and Your Genes

## Steven J. Heine

**W. W. NORTON & COMPANY**
*Independent Publishers Since 1923*
New York | London

For information about permission to reproduce selections from this book,
write to Permissions, W. W. Norton & Company, Inc.,
500 Fifth Avenue, New York, NY 10110

For information about special discounts for bulk purchases, please contact
W. W. Norton Special Sales at specialsales@wwnorton.com or 800-233-4830

Manufacturing by LSC Communications, Harrisonburg
Book design by Lovedog Studio
Production manager: Julia Druskin

Library of Congress Cataloging-in-Publication Data

Names: Heine, Steven J., author.
Title: DNA is not destiny : the remarkable, completely misunderstood
    relationship between you and your genes / Steven J. Heine.
Description: First edition. | New York : W. W. Norton & Company, [2017] |
    Includes bibliographical references and index.
Identifiers: LCCN 2017000693 | ISBN 9780393244083 (hardcover)
Subjects: LCSH: Genes—Popular works. | Human genetics—Popular works. |
    Molecular genetics—Popular works. | Genetic engineering—Popular works.
Classification: LCC QH447 .H45 2017 | DDC 572.8/6—dc23
LC record available at https://lccn.loc.gov/2017000693

ISBN 978-0-393-35580-2 pbk.

W. W. Norton & Company, Inc.
500 Fifth Avenue, New York, N.Y. 10110
www.wwnorton.com

W. W. Norton & Company Ltd.
15 Carlisle Street, London  W1D 3BS

1 2 3 4 5 6 7 8 9 0

*To my parents.*

*Thanks for the genes and the experiences.*

# Contents

# 1.

# Introduction

**IN THE SECOND WEEK OF APRIL 2003, OUR** world changed forever: we sequenced the first complete human genome.

We now have access to information about ourselves that no previous generation has ever had: we can peer directly into our own genetic makeup. We each have a unique string of nucleotides in our cells that contributes to who we are, and since that fateful week in 2003, we are able to unravel this string and read it. It's a 6-billion-letter autobiographical code that seemingly flows straight from the pen of God. Our genome contains deep secrets about us: where our ancestors came from, which diseases we'll likely avoid and which ones might kill us, and what kinds of physical and psychological attributes we're predisposed to have. This remarkable scientific revolution seems to provide us with nothing short of a window into our souls.

It is hard not to be both excited and frightened about the vast potential of this achievement. Our knowledge of the human genome promises a flood of great medical advances. For example, President Bill Clinton foresaw a future in which genetic discoveries made it "conceivable that our children's children will know the term 'cancer' only as a constellation of stars."[1] And we may someday have the potential not only to treat diseases, but also, via genetic engineering, to eliminate hundreds of genetic diseases for good. Many talk breathlessly about a future with designer babies, where prospective generations will not only be born with fewer genetic vulnerabilities to disease, but also with genetic enhancements that make them more fit and more intelligent than ever before.[2]

Moreover, our knowledge of the human genome can be used as a nearly ironclad means to identify individuals. For example, the popular crime drama *CSI: Crime Scene Investigation* was based on the premise that crime scenes were riddled with invisible traces of DNA that unambiguously identified all who were involved. Indeed, this plays out in real court cases that depend on DNA testing to identify the perpetrators. Sometimes this endeavor can get taken to absurd ends, such as the Hong Kong Cleanup Initiative, which promises to collect DNA traces from litter to create a picture of the litterer's face, to publicly identify and shame him or her. The same logic has been exploited by PooPrints DNA testing kits, which promise to identify delinquent dog owners who have failed to clean up after their pet.[3]

Our genes can also tell us long lost tales about who our ancestors were, and where they likely came from, based on similarities between our own genomes and those of people living around the world. And the ability of genome sequencing to identify people's ancestry and personal history has even been extended to those long since gone; for example, scientists have read the genetic script of a Neanderthal man who lived approximately 38,000 years ago, from DNA in his remains. Our genes seem to stand as a nearly permanent record of our life and the lives of our predecessors.

The genomic revolution promises to completely upend our understanding of the world. And with the advent of direct-to-consumer genetic testing companies, such as 23andMe, you can now affordably get yourself genotyped. What will you do when you learn about the genetic secrets of your own life? My bet is that you're going to respond in all the wrong ways.

This is not because of any specific genes that you might learn about, but because of the psychological machinery in your brain which influences *how you think about genes.* We come equipped with a set of psychological biases that ascribe almost mystical powers to our genes. These biases can lead to racism, sexism, or eugenics, but they can also foster tolerance for others, sympathy, and potential for progress. We're going to explore how these psychological biases work, how they make us vulnerable to the massive hype surrounding the genomic revolution, and how we can corral them into more effective ways of thinking.

## The Genomics Revolution and You

If you feel somewhat overwhelmed and anxious about this genomic revolution, you're not alone. Technological innovations can always be challenging to cope with, but the problem is even more daunting when it comes to genomics. One reason we find genomic progress so disturbing is because of its unprecedented blinding speed. Much has been made of the breakneck rate of progress underlying the personal computer revolution: Intel cofounder Gordon Moore's famous "Moore's Law" claimed that the number of transistors on microchips would double every two years. But the pace of the genomic revolution far outstrips this. When the first complete human genome was sequenced in 2003, it cost a few billion dollars and thousands of person-years worth of efforts. Yet barely a decade later, a complete sequence of your own genome was available for about $1,000, and would be ready for you within a matter of days. Genomic sequencing is no longer a special tool for those pursuing research questions in large international scientific consortiums; it is now available to you as a consumer product and could soon be part of your family's routine medical care. In a seeming blink of an eye, the genomic revolution has infiltrated our lives.

Another reason why people find the genomic revolution to be disturbing is that, unlike other scientific revolutions, this one is personal. Splitting the atom may have changed the world, but discovering your own genome will change how you view yourself. Reading over your unique sequence of DNA can feel as though you are gazing into a crystal ball, discovering secrets that have been passed down to you from your ancestors.

But the main reason we are so anxious about the genomic revolution is that we are psychologically equipped to *misunderstand* it. Unlike, say, the study of subatomic physics, where almost no one outside of the physics community feels that he or she can make heads or tails of it, the notion that we possess genes that make us who we are makes intuitive sense. But it turns out that conclusion is inaccurate, or at least imprecise. Yet we persist in this belief that our genes control our lives. We are *genetic fatalists*.

To see our fatalistic thoughts on genes in action, let's visit the 2014 television series, *Dead Famous DNA*, which was broadcast on Britain's Channel 4. The premise of the show is described by its host, Mark Evans: "I want to hunt down the DNA of the most famous people who ever lived. The aim is to find out who they really are." Evans enters the seedy world of body-part trafficking to obtain and genotype parts of dead celebrities and reveal the secrets of their lives. The show aspires "to understand what made Einstein so intelligent, Marilyn Monroe so attractive, and Adolf Hitler so evil." The answers to these fundamental questions were supposedly all waiting to be read directly from the genomes of these famous people.

My favorite episode was when Evans sought to explain what killed Elvis Presley. He obtained a hair sample, which was purportedly snipped from Elvis's head, and placed in safe-keeping by his barber. (Authentic samples are a key challenge for the show: in this same episode Evans paid $5,000 for what was purported to be a lock of King George's hair, only to have the genotyping lab inform him that he had just bought a piece of a very expensive wig.) Evans was confident this hair sample must have come from Elvis, because genetic testing revealed risk factors for migraines, glaucoma, and obesity; all of which are conditions that Elvis suffered from.[4] And the purported smoking gun was found at the precise genetic address RSID 193922380, which is located on the MYBPC3 gene that sits on chromosome 11. At this location, the sample possessed a G nucleotide (in contrast to a C, like most people have). This particular genetic variant was suspected to be associated with familial hypertrophic cardiomyopathy, a serious kind of heart disease, and Elvis did indeed die of a heart attack. Evans inquired about this variant to geneticist Stephen Kingsmore. "Do you think this is significant enough to tip it towards . . . ?" Kingsmore replied in somewhat guarded terms: "It looks pretty suspicious that this may indeed have contributed potentially to Elvis's death." This was summarized by Evans as evidence that Elvis's "early death was his genetic destiny," and was further embellished by headlines published around the world, such as that of *The Mirror*, "Shocking DNA results reveal Elvis Presley was always destined to die young." The mystery surrounding Elvis's early death had apparently been solved.

Yet if we look more closely at this claim, it starts to fall apart. The variant of Elvis's that the show discussed, having a G nucleotide at RSID 193922380, has not at all been shown to be a strong predictor of familial hypertrophic cardiomyopathy.[5] To make matters worse, there were no signs from Elvis's autopsy to suggest that he could ever have been diagnosed with familial hypertrophic cardiomyopathy.[6] It's true that, near the end of Elvis's life, he was indeed suffering from migraines and glaucoma, and he had become obese. But did the genetic variants in his alleged DNA cause these? Most certainly not in any direct way. In the case of obesity, there are at least 97 common genetic variants that increase the likelihood of becoming obese.[7] Virtually all of us have several of these variants, and the one obese-relevant variant identified in Elvis's DNA by Dr. Kingsmore is not even one of the stronger predictors. The same goes for Elvis's genetic risks for migraines[8] and glaucoma;[9] neither of the genetic variants identified in Elvis's DNA are strong predictors for these. Saying that these genes caused him to have these conditions is akin to saying that the unusually warm winter was what caused Donald Trump to win the Republican nomination for president in 2016. The weather may influence voter turnout, but it certainly isn't the deciding factor in the outcome. There are just far too many other factors involved.

But the more obvious question that we should be asking is whether the best way to solve the mystery of what caused Elvis's heart attack is to look at his genome in the first place. After all, Elvis was hardly a poster child for a healthy lifestyle. Near the end of his life, he was reportedly addicted to multiple drugs, including Demerol, and had been hospitalized numerous times for overdosing on barbiturates.[10] His autopsy report identified 11 drugs present in his system at the time of his death.[11] Moreover, the excess of Elvis's diet has become the stuff of legend, with his alleged favorite sandwich, the Fool's Gold Loaf, consisting of a whole loaf of French bread, hollowed out and filled with an entire jar of peanut butter, another jar of grape jelly, and a pound of bacon. The BBC documentary series *Arena* estimated that near the end of his life, Elvis's daily caloric intake rivaled that of an Asian elephant![12] If these legends carry any truth, it would make more sense to inquire about the genetic variants that Elvis possessed that kept him so relatively thin in the face of his obscenely gluttonous diet.

The claim that Elvis's death was telegraphed in his DNA feels like a more satisfying explanation than saying he was done in by an excess of greasy sandwiches, though, doesn't it? But that account, however well it may fit with the fatalistic ways that we tend to think about genes, is a preposterous misreading of the predictive strength of Elvis's genes. The popularity of *Dead Famous DNA* is a perfect example of the way our psychological biases about genes have run amok.

## The Psychology of Genetics

Unfortunately, the fatalistic attitude toward genes that is on display in *Dead Famous DNA* is not at all unusual. Why do ideas about genes affect us in this way? To address this question, I launched a series of psychological experiments with the help of my graduate students to explore what happens when people encounter genetic arguments in their lives. Do people ignore this information? Treat it the same as other kinds of information? Or do they give it special attention?

In one of our studies, we wanted to see how people would respond to learning about different causes of obesity.[13] We had Canadian university students come into our lab to read some newspaper articles and answer some questions about them. First they read some distractor articles to throw them off the scent of the true purpose of our study. Then, one group of students who were randomly assigned to a "genetics" condition read an article about scientific research regarding "obesity genes"—genes that affect how much people weigh. Another group assigned to a "social experiences" condition read about scientific research that showed that the weight of people's friends affected how much they weigh. A third group assigned to a "control" condition read an article that was unrelated to obesity—it was about corn production. These were the kinds of news articles that you might encounter with your morning paper, and they were all based on legitimate scientific research. Later on, the students were told they would be participating in a second study that involved food preferences. They were provided with a bowl of cookies and were asked to evaluate the taste of them on various dimensions. This last bit was a ruse, as actually the whole point of our study was to see how many cookies people ate after reading the newspaper articles. What did we find? Those who read about the existence of obesity genes ate one-third more cookies than those who read either about social causes of obesity or who didn't read anything about obesity. Even though our weight is influenced by both our environments and our genes, it was only the arguments about genes that affected people's behaviors. Learning about the existence of obesity genes led people to act in ways more likely to make them obese!

A key reason why we find genes so compelling is that we have a particular preference for explaining the events that unfold around us. In every society that has been investigated, there is clear evidence to show that we are predisposed to think of the world as emerging from hidden underlying *essences*. We believe families share traits because of their common "blood"; Indian yogis derive their power from some hidden prana energy; traditional Chinese medical practitioners diagnose someone with, for instance, "liver fire," on the basis of having an imbalance between his or her yin and yang; medieval alchemists sought the mystical forces of a philosopher's stone with the hope of changing lead into gold; and Yoda was able to perform his magical Jedi feats because of his command of "the Force." With the advent of genomic science, we now have a new essence: our genes. And when we think of genes, we activate this ancient and universal human tendency to imagine essences, and we are left with an unshakable sense that genes determine life outcomes.

We can see our essence-based conception of genes in everyday discourse. For example, President Obama said that racism is "part of our DNA";[14] Pink sings that her love for partying "is genetic";[15] Brad Pitt argued that "it's in our DNA" as Americans to own a gun;[16] Donald Trump attributes his drive for success to having "a certain gene";[17] and so many corporations claim that innovation is part of their DNA that there's a growing backlash against the clichéd expression.[18] These statements reveal that we not only regularly stud our conversations with genetic concepts, but when we do, we also equate our DNA with some inherent and unchangeable essence that makes us who we are. Our daily conversations about genetic concepts reinforce the idea that our genes are our fate.

And there is a dark side to these so-called essences. Imagined essences underlie why some people think that certain races are inferior to others, that women should be treated differently than men, that gay men and lesbians are ultimately different from heterosexuals, that people with mental illnesses are dangerous, or that criminals won't ever be able to escape a life of crime.

These dangerous associations imbue genetic information with an unusually powerful, almost sinister feel. Reflecting these concerns, when the Human Genome Project was first established as a joint venture between the Department of Energy and the National Institutes of Health in 1990, there was already much discussion about the potential dangers that genetic information entailed. In response to these fears, a special program was established called the Ethical, Legal, and Social Issues Research Program. It was mandated by law that a total of 5 percent of all funding for human genome research would be set aside for this program.[19] Given that the sequencing of the genome ultimately totaled in the billions of dollars, this mandatory budget of 5 percent for research on implications of the genome was an unusually enormous sum for a research program that fell largely within the humanities and social sciences.

The fears we have about genetics are not necessarily misplaced; when we talk about genetics, we can quickly get into some rather treacherous topics. For example, research on human genetics in the first half of the 20th century was deeply connected with the study of eugenics. At that time, if you were a researcher studying human genetics, it's quite likely that you also self-identified as a eugenicist, given how closely the fields were linked back then.[20] Although the scientific study of eugenics was hastily abandoned following the rise of the Nazis and the Holocaust, today there are again new concerns about how genetic innovations may be applied in disturbing ways.

For example, in 2013, Senator Rand Paul gave a speech about future genetic nightmares where he described events from the dystopian movie *Gattaca*. Although Paul's speech got more media coverage because it turned out that he had plagiarized it from the movie's Wikipedia page, his speech is also noteworthy for his assumptions that genetic research stands to usher in a society right out of *Brave New World*, where "the state select(s) for perfection." Paul asked, "Will we have the strength of character to resist a world where eugenics is practiced voluntarily? Are we prepared to select out the imperfect among us? In the process will we eliminate some part of our humanness, our specialness, when we seek perfection?"[21]

With the advent of new genetic engineering technologies such as CRISPR/Cas9, concerns about the proliferation of designer babies,[22] contentious political debates about the perceived dangers of genetically modified organisms, and the rise of direct-to-consumer genomics companies such as 23andMe, our concerns about dystopian genetic futures will only become more common.

The genomics revolution promises to change our lives in many exciting ways, but along with these genuine scientific advances, the revolution also comes packaged with an enormous amount of unwarranted hype. Our genomes provide us with a grand vista from which to view ourselves and our world, but as long as we're peering through a set of distorted lenses, we're going to see things that aren't really there. Over the next several chapters, we're going to take a closer look at our mistaken assumptions about our genes, and gain a clearer understanding of how our genes actually influence us. We'll discuss my own experiences getting genotyped, and how my own psychological biases interfered with how I made sense of the feedback I received. We'll see that although discussions about our genes can arouse our darkest fears and prejudices, a closer look reveals in sharp relief that our DNA is not our destiny.

## 2.

# How Genes Make You Who You Are

**IT WAS A BAD CASE OF TEST ANXIETY THAT** launched the field of genetics. A brilliant young student, Johann Mendel, aspired to become a schoolteacher near the town of Heinzendorf in Silesia (now Hynčice in the Czech Republic), where he had grown up. His family had great faith in his promise as a scholar, and his sister Theresia had loaned him her portion of the family estate to pay for his studies. But despite this support and encouragement, and his obvious brilliance, Mendel suffered an enormous anxiety attack during his oral examinations and failed the certification exam that would have qualified him as a teacher. Six years later, he gave up in the middle of his second attempt at the exam. Humiliated, he resigned himself to life as a monk at St. Thomas's Abbey in the nearby city of Brünn (now Brno), where he took on the name of Gregor. To Mendel's surprise, the monastery turned out to be an ideal place for his scientific studies because he had few distractions, access to the voluminous monastic library, and the support of the monastery's abbot, Cyrill Napp, who recognized that Mendel's skills for scientific study clearly outweighed his talents as a priest. Indeed, Napp was so taken by the passion and discipline of young Gregor that he ordered a new greenhouse be constructed for him to pursue his scientific interests. And it was here, in the gardens and greenhouse of the St. Thomas monastery, that Mendel conducted one of the most famous series of experiments in the history of science. He grew peas.[1]

Mendel sought to provide a quantitative foundation for one of the most poorly understood scientific questions. Why do offspring resemble their parents? It was so clear that they did, but the mechanisms behind this remained unknown. Although Mendel had earlier attempted to address this by breeding mice, the bishop had deemed it inappropriate for a celibate priest to be studying mice having sex, so he turned to breeding plants instead. He worked with a number of different species, but it was his eight-year experiment with peas, *Pisum sativum*, that earned him his later fame. During this period, Mendel would brush the pollen from one variety of pea onto another, about 29,000 times, thereby cross-fertilizing the varieties. He then recorded information about their offspring. The plants differed from each other across seven different traits: seed shape (either wrinkled or round), location of the flowers (at the tip or along the stem), seed color (green or yellow), seed coat (white or gray), shape of ripe pod (either inflated or pinched), unripe pod color (green or yellow), and height (tall or short). Mendel wanted to see what happened when you crossed, for example, a pea with a green pod with one with a yellow pod. You might expect that the colors would blend together to create a pod of intermediate color—say, one that was chartreuse. But he never found this; when he crossed the green peas with yellow peas, the resulting second generation of peas was all yellow. Mendel then crossed the peas of this second generation with each other: yellow peas with other yellow peas. In doing this, and counting very carefully, he found that the subsequent generation produced peas that were either yellow or green, and with the highly specific proportion of three yellow peas for every one green one. This same ratio emerged for all of the seven different traits that he had studied, and it pointed to something profound that forever altered the way scientists understood heredity.

Mendel's experiments revealed that each plant inherited one "character," as he called them, from each of their two parents. Further, he identified that these characters came in one of two forms; they were either dominant or recessive. Recessive characters (such as the ones that create green pods) were only visible in the next generation if the offspring inherited the same recessive characters from both parents; otherwise, the next generation only displayed the dominant character (such as the ones that create yellow pods). Mendel signified these characters with a capital letter, A, if they were dominant, and a lower case letter, a, if they were recessive. This led to four different combinations in the offspring that Mendel's diligent counting revealed were equally likely: AA, Aa, aA, and aa. This labeling is still used today, and the characters that they refer to are now known as genes. These genes come in different versions (such as whether the gene is associated with a yellow or green pod), which are termed alleles.

It was in the garden of St. Thomas that Mendel had discovered the quantum nature of inheritance. Genes, and their corresponding traits, never blend—our parents' blood does not get mixed together—rather, genes segregate such that we inherit one copy of the gene from each parent. The genes always remain intact from generation to generation, like pearls on a string, and each parent passes down a particular collection of these pearls to their children. The specific combination of these pearls that we receive from our parents (that is, the alleles such as Aa or aa) are termed our genotypes. Although genotypes remain hidden from our view, they specify the kinds of proteins that are created in an organism. The set of observable characteristics of an organism, which are products of the proteins created by genotypes, is called a phenotype (e.g., whether the pea pod is actually green or yellow). This discovery laid the foundation of modern genetics.

Mendel realized from his results that he was onto something big, and he published his findings in 1866 in a 44-page article in the *Proceedings of the Natural History Society of Brünn*, a journal that was carried in the best libraries at the time. He also sent out 40 copies of his article to several leading scientists around the world. But the response that he got was virtual silence; few people read his article at the time, and of those who did, none of them seemed to fully appreciate what Mendel had found. Charles Darwin was one of the scientists to whom Mendel sent a copy of his article, and he received it at a fortuitous moment. Darwin was being challenged by a Scottish engineer named Fleeming Jenkin, who pointed out that evolution could not possibly occur if traits blended together. A gazelle, say, that could run a little bit faster than the other gazelles would indeed be more likely to survive and pass on its traits to its offspring, but its offspring would inherit a running speed that was a mixture of one fleet parent and one slow one, thereby diluting the advantageous trait until it would no longer carry any selective advantage. Mendel's model of quantum inheritance showed that traits don't dilute over generations, and his findings served as a direct answer to Jenkin's criticism. Alas, it seems that Darwin never read the article that Mendel sent to him.

Mendel lived for another 18 years after the publication of his article, and he continued breeding experiments with a number of other plant species as well as bees, but he never again found the same kind of clear evidence for quantum inheritance that he had found with peas. He later had to give up his research completely when he became abbot of St. Thomas. At the time of his death, Mendel's research remained virtually unknown to science. It wasn't until 16 years after he had died that his work was rediscovered and he was christened the father of genetics. The posthumous rediscovery of Mendel's work, and his promotion to the pantheon of science, has made Mendel's life story a beacon of hope for all scientists who feel that their work is underappreciated.

## DNA and the Making of Proteins

Our genes are stored in 23 pairs of chromosomes that reside inside the nuclei of almost all the cells in our bodies. When sperm and egg cells are produced, the 23 pairs of chromosomes are reduced by a process of meiosis into 23 single chromosomes in each sperm or egg cell. Each set of single chromosomes contains a unique shuffling and recombining of the genes that were contained in each original chromosomal pair. Like poker players, sperm and egg cells are basically dealt a unique set of genes from the deck of the complete set of chromosomes in your body. This is why siblings (aside from identical twins) are never duplicates of each other. On average, we inherit approximately half of our genetic code from each parent, yet it's a different combination for each sibling, so that siblings share about half of their unique genetic markers with each other. One of those 23 pairs of chromosomes determines our sex. Typically, women inherit an X chromosome from each parent, so they have a pair of X chromosomes. Men, in contrast, typically inherit an X chromosome from their mother and a Y chromosome from their father.

Each of our chromosomes is a long unbroken string of deoxyribonucleic acid (DNA), which consists of two long polymer molecules that spiral around each other. Binding these two molecular strings together in a double helix are a series of the basic building blocks of DNA: the nucleotides, which are organic molecules that come in four different flavors (guanine, adenine, thymine, and cytosine), and are identified by the letters, G, A, T, and C. Each nucleotide is matched with its corresponding "spouse" on the other strand of DNA, such that G is always matched with C, and A is always matched with T. There are about 3 billion monogamous pairs of these nucleotides total in each of our cells. Just like the letters of the alphabet, which can be arranged into a nearly infinite set of possible words, these four nucleotides can be arranged into a nearly infinite set of possible combinations, typically several thousands of nucleotides long, which form our genes.

Genes function by leading to the "expression" or creation of proteins. When a gene is expressed, the double helix of DNA will unravel, much like a zipper unzipping, and will replicate a mirror image of itself on a strand of messenger RNA, a similar long polymer molecule that binds to the string of nucleotides of DNA. This messenger RNA then detaches from the DNA molecule and moves to another part of the cell to a ribosome, which is essentially a protein factory. There, the messenger RNA is translated into a unique protein. Amazingly, all the different proteins in our bodies, from the neurochemicals that course through our brains to the enamel in our teeth, come from the four nucleotides that constitute our DNA. It's the particular combination of the letters that determines the nature of the protein. In this way, DNA is very much a molecule of *information*, not unlike a string of computer code that carries instructions for how proteins can be made.

We have approximately 21,000 protein-coding genes lined up along our chromosomes; a number roughly similar to that of other mammals, yet curiously smaller than that of many plants, such as tomatoes, which have more than 30,000 genes.[2] This may not seem like a lot of genes to work with, especially since we share so many of the same genes with other species (we share approximately 98 percent of our genes with chimpanzees, and 92 percent with mice; even the lowly yeast shares about one-quarter of our genes).[3] Before the results of the Human Genome Project were announced in 2003, geneticists had always assumed that the actual number of human genes would be much higher. The reason that we can get by with so few genes, and that we don't look anything like a mouse despite sharing 92 percent of its genes, is that each gene we possess is remarkably versatile, and the process by which genes express proteins is extraordinarily complex. First, a single gene can produce several different but related proteins, so the total number of proteins in our bodies is many times greater than the number of genes we have. Moreover, each of our phenotypic traits, such as our nose shape, or thickness of hair, is typically the result of many different genes interacting; there is a great deal of redundancy in the system, and any single gene usually has only a small influence on associated phenotypes. In addition, our genes are expressed differently over time; molecular signals, often in response to specific environmental events, determine precisely when and where a particular gene will be expressed. On average, each of our proteins lasts about 80 days before it is replaced when our genes express new proteins.[4]

This copying of DNA to RNA to proteins is extremely precise, and there are various mechanisms that ensure that the copying is done with virtually no mistakes—imagine the world's best secretary at work. But given the sheer amount of copying that is done, very occasionally errors are produced. If these errors are rare enough that they are shared by only a small percent of the population (i.e., less than 1 percent), they are termed mutations, and when the errors have become more common throughout a population (i.e., greater than 1 percent), they are called polymorphisms. One common error is akin to a typo—you might accidentally type a C instead of a G. These errors involve only a single nucleotide, and when they are relatively common in a population, they are called single-nucleotide polymorphisms, or SNPs for short. Structural errors occur as well, like the kind you might make when using the cut, copy, and paste commands in your word processor. Strings of DNA are deleted, or inserted where they don't belong, or even repeated again and again, like CAGCAGCAGCAG.

It is by the virtue of these occasional copying errors that every individual is genetically unique; each human newborn averages about 60 new mutations throughout his or her DNA.[5] The vast majority of these mutations have no discernible impact on the proteins that are produced. However, some mutations do affect the nature of a protein, and these can ultimately have an impact on the child's phenotype. Very rarely, the mutated gene is an improvement on the original one, and if so, it will become more common in future generations as the result of natural selection. For example, several thousands of years ago, on the SLC24A5 gene, at location rs1426654, some individuals were born with an A nucleotide, which was a miscopy from the more common G.[6] The A allele contributes to a lighter skin color, and those humans who were living in northern latitudes who were born with the A allele were able to absorb more sunlight, which allowed them to synthesize more vitamin D. This was advantageous in the northern latitudes where direct sunlight is scarcer, and those who possessed this polymorphism were more likely to have surviving offspring than those who possessed the G allele. In contrast, the G allele is more adaptive at latitudes where too much sunlight can create problems, and the G allele remains more common there.

More commonly, a mutated gene bears a cost to the individual, and if the costs are significant enough, the mutations will become less common in future generations because the individuals with those mutations are less likely to survive and reproduce.

The ways that DNA gets translated into proteins are further complicated by epigenetic influences. Epigenetics refers to ways that proteins are expressed through molecular processes that are separate from the instructions contained within the DNA sequence itself. Remarkably, these molecular processes can be triggered by our experiences and our environment. The best understood kind of epigenetic influence is through DNA methylation, in which a methyl group (a molecule composed of one carbon and three hydrogen atoms) binds to the nucleotides of the DNA itself, and thereby controls how often that gene is expressed. Incredibly, these methyl groups can be heritable; they can be copied through meiosis and be inherited by the next generation. This means that our life experiences do not only affect us, but can even affect our children.

A dramatic example of this was recently found in an experiment with mice: researchers trained a group of mice to become afraid of the odor of acetophenone (which smells like cherry blossoms) by regularly pairing an electric shock to their feet with the odor.[7] Soon, the mice showed a startle reaction whenever they encountered that specific odor. These mice were then allowed to breed, and their offspring also showed the same startle reaction to the smell of cherry blossoms. These offspring were also bred and the next generation continued to show this same cherry-blossom-specific fear. The researchers dissected the animals' brains and found that the mice had larger receptors that responded to this particular smell. Moreover, an analysis of the DNA of the sperm of these conditioned mice found that there were different methylation patterns in the sperm of the mice who learned this fear compared with other mice. To rule out the possibility that the mice had learned this fear from their parents, the researchers took the sperm from males who were trained to fear the smell of cherry blossoms and artificially inseminated female mice in another lab. The brains of these offspring were dissected, and they also showed the same larger receptors for this smell. A specific phobia, learned by one generation, was passed through the sperm to other generations, not through the DNA sequence itself, but through the methylation patterns that affected the gene's expression. In other words, one generation's experiences led to heritable changes. We have much more to learn about epigenetic inheritance before we can draw solid conclusions, but findings such as these are fascinating.

## From Genotypes to Phenotypes

Phenotypes are ultimately what matter in terms of our lives—they are what characterize our bodies and our behaviors. Fast gazelles are more likely to escape predators than slow ones, tall pea plants can avoid the shade better than short ones, and children with symptoms of attention deficit disorder struggle more to pay attention in class than those without the symptoms. Genotypes shape all of these. The problem, though, of identifying how genotypes get translated into phenotypes is not that straightforward.

The father of behavioral genetics, Sir Francis Galton, first bumped up against this problem in 1869, when he became fascinated by the idea that genius seemed to pass through families. It's certainly not surprising that Galton had this intuition; sitting at the table at his family's Christmas dinner, he must have noticed how his own genius was matched both by his half-cousin, Charles Darwin, and their eminent polymath grandfather, Erasmus Darwin. One possibility for this concentration of intellect in the Darwin–Galton family is that Grandfather Erasmus had provided a stimulating educational environment that had cultivated this blossoming of talent. But Galton was convinced of a different kind of explanation—he believed that there was a biological essence underlying genius. He had the brilliant inspiration that he could provide evidence for this by studying twins. Galton recognized that there were two different kinds of twins— those that were no more similar to each other than other siblings, and those, strongly alike, "due to two germinal spots in the single ovum." He sent letters to 35 pairs of seemingly identical twins and another to 20 pairs of apparently fraternal twins to look for evidence of any similarities in their behaviors. The replies that he received revealed that his fraternal twins indicated no particular degrees of similarity beyond that of any other pairs of similarly aged siblings. In sharp contrast, many of the replies from his identical twins were striking in how alike they were. One pair of identical twins developed a toothache at the same time, in the same tooth, which needed to get extracted; another pair developed the same curious difficulty at the age of 20, so that they could no longer run down stairs. Identical twins reported having the same dreams at the same times, unknowingly buying each other the same gifts, or getting the same jobs. The similarities of the identical twins were so striking, and indicated that the role of genes was so powerful, that Galton claimed: "My only fear is that my evidence seems to prove too much and may be discredited on that account, as it seems contrary to all experience that nurture should go for so little."[8]

Galton's curious findings with twins demonstrated that the degree of biological relatedness between individuals would allow us to determine how much of any given trait was *heritable*—that is, genes play a role in explaining why one individual is different from another—and this led to the founding of the field of behavioral genetics. The two most common methods of estimating heritability are to compare identical and fraternal twins, or to compare adopted and biological children. Because identical twins share 100 percent of their genetic variants, whereas fraternal twins only share 50 percent of theirs, whenever we see evidence that identical twins are more similar to each other than fraternal twins for a given trait, we have evidence that the trait is heritable. Likewise, if children are more similar to their biological parents (who are 50 percent genetically similar) for a given trait than children are to their adopted parents (who are close to 0 percent genetically similar), then we again have evidence that the trait is heritable.

What exactly does it mean to say that a trait is heritable? This is not at all an intuitive concept, so let's unpack it a bit. Height is a trait that is highly heritable: twin and adoptee studies indicate the heritability of height is as high as 80 percent to 90 percent.[9] So what does this mean? In contrast to what many people intuitively think, heritability does *not* refer to the degree that something is inherited from one's parents. So, it does *not* mean that 80 percent to 90 percent of your height was inherited from your parents' genes, while the remaining 10 percent to 20 percent came from your experiences. Rather, heritability refers to how much of a trait's *variance* within a given sample is due to genetics. For example, a heritability for height of 80 percent means that among individuals tested *within the same sample*, about 80 percent of the reasons why some of those people are taller than others is due to their respective genes. The remaining 20 percent is due to experiences in their lives (either while developing in the womb, or after that in their experiences growing up). Importantly, heritability estimates are *always* limited to the samples that were tested—they can't speak to people who live in different circumstances. As we'll discuss later, this point has important consequences when we look at a single trait across separate groups.

So what kinds of traits are heritable? Given what I just said about height, you probably wouldn't be surprised to learn that other physical characteristics, such as eye color, body mass, or metabolism rates are heritable. What about psychological characteristics? Many of these are too. Research reveals that there are strong genetic contributions to such traits as intelligence, personality, self-esteem, and the likelihood of developing schizophrenia. Actually, evidence for the genetic foundation of our minds is far broader than this—there's evidence for heritability of whether you believe in God, are racist, whether you like jazz, how much television you watch, whether you support euthanasia, were a bully as a child, are a compulsive hoarder, volunteer to donate blood, spend more money on chocolate than batteries, how happy you are, or how much you enjoy science fiction books.[10] There is even evidence for heritability of the likelihood that you'll get mugged![11] It seems impossible that the genes coiled up in our cells could somehow leave us standing alone on a dark street, shouting in vain as a hooded stranger bolts away with our wallet in hand. But the evidence of the heritability of these human traits and behaviors from behavioral genetics studies is strong, despite how counterintuitive it all seems.[12]

In fact, there is such an enormously broad body of evidence for heritability in our behaviors that the behavioral geneticist Eric Turkheimer offered what he called the "first law of behavioral genetics," which states that *all* human behavioral traits are heritable.[13] That is, everything that people do is more similar between identical twins than it is between fraternal twins. This "law" is recognized as somewhat of an overstatement. For example, the particular religion you come to believe in (e.g., Baptist, Zen Buddhism, Islam) is not heritable (although the depth of your religious commitment is).[14] Rather, what this law means is that the evidence for heritability is so vastly extensive that we assume by default that a behavior is heritable until proven otherwise. And remarkable as it may seem, almost all behaviors do indeed seem to be tied, in some ways, to the genes we're born with.[15]

Studies of twins and adoptees only provide general evidence that genes contribute to phenotypic variability; that is, they show that some variation in height, for example, is partially due to our genes. And for most of the first century of behavioral genetics, that is really the only kind of answer that the field could provide. But recent technological advances have finally enabled us to identify the particular genes associated with different traits. There are a variety of ways that gene associations can now be identified (such as linkage studies, which we won't get into here), and the following are two of the most commonly used approaches. First, researchers can have a specific gene in mind, and they can investigate whether variants of that candidate gene are more common in people with a certain condition, such as Crohn's disease, than in those without it. In these candidate gene studies, the particular genes are chosen based on some hypothesized relation to the condition. In the case of Crohn's disease, several previous studies had pointed to a particular stretch of chromosome 16 as likely being involved with the disease. The NOD2 gene was in the right location, so researchers chose to investigate it. Indeed, they found that people with a particular variant of the NOD2 gene were three times more likely to develop Crohn's disease than those without it.[16] To use a military analogy, candidate gene studies are like the Seal Team 6 operation to assassinate Osama bin Laden. Based on their intelligence reports, the US government suspected that bin Laden was living in a three-story compound at the end of a dirt road on the outskirts of the city of Abbottabad, Pakistan. The United States then launched a surgical attack on the suspected compound with the hope that bin Laden was indeed there, and in this case, the results of their attack confirmed the hypothesis.

If candidate gene studies are akin to surgical military strikes, then another common means for identifying relevant genes, genome-wide association studies, are the equivalent of the carpet bombing of Dresden. These studies rely on a DNA chip that can quickly and cheaply provide genotype information for several hundred thousand different SNPs: those specific places along the genome where individuals differ by a single nucleotide. Genome-wide association studies collect genomes from people with a condition such as Crohn's disease, and from people without it, and then compare *all* of the two groups' SNPs; after bombarding every possible statistical association, the researchers can search through the wreckage to see what was found. The studies do not depend on any prior knowledge or hypotheses about what genes might be involved; rather, through brute statistical number-crunching, the researchers identify which of the hundreds of thousands of SNPs are more common in people with the condition than those without it. In the case of Crohn's disease, more than 70 different SNPs have been identified that each predicts the disease.[17] Genome-wide association studies have provided a virtual flood of data, and they've been used to identify genetic variants in all kinds of conditions, from having blond hair to being diagnosed with prostate cancer. After a genome-wide association study has been conducted, later studies will then try to see if the previously identified SNPs can be found again in new samples. Given that this method is guided solely by statistics rather than by theory, often it turns out that many of the SNPs tagged in an initial genome-wide association study are later found out to have emerged just as a result of chance. The goal is that over the course of several of these studies, a set of SNPs will emerge reliably, and researchers will be able to conclude that those variants play a role in the development of the condition. Behavioral genetics thus typically relies on studies of twins and adoptees to identify which conditions have a genetic basis, and then turns to various kinds of gene-association studies to identify which specific genes are involved.

## Switch-Thinking and the Evidence for Genetic Determinism

The Mütter Museum of medical curiosities in Philadelphia is not for the fainthearted. Its dimly lit glass cabinets contain a collection of disturbing specimens of all the many horrible ways that the human body can go awry. There are displays of an impossibly swollen nine-foot-long colon, wax models of skulls with horns growing out of them, and the death cast of the bodies of the original Siamese twins, Chang and Eng. But it is not until you descend the dark staircase to the basement of the museum that you encounter one of the museum's most fascinating attractions—the skeleton of Harry Eastlack that hangs eerily in the corner. As you get closer, you realize it's unlike any other skeleton that you've ever seen. The skeleton is twisted and contorted, and you can see that the individual bones are fused together to form a bone shell, not unlike that of a tortoise.

The first sign that there was anything untoward in Eastlack's life appeared at the age of four, when he broke his femur while playing with his sister. The fracture didn't heal properly, and shortly afterwards his knee and hip began to stiffen. When examining him, the doctors were surprised to find bony deposits growing on the muscles of his thighs, which they tried to surgically remove, but the bone only grew back thicker, and more extensively, than before. A few years later, he accidentally bumped his buttocks into a radiator, and a new bony growth began to form where the bruise had been. From that point on, Eastlack's life became an endless series of futile surgeries; the doctors seemed more like gardeners as they cleared away in vain whatever problematic bony deposits they could find along his body, only to watch them grow back with a vengeance, like a tangle of uncontrollable weeds. By his mid-twenties, Eastlack's vertebrae had fused together as his back muscles were replaced with solid bone, and his neck was left permanently jutting off to the side. His breathing became more difficult because his lungs could no longer fully expand, and after his jaw too became encased in a cast of bone, Eastlack was no longer able to feed himself and could only speak through his clenched teeth. Six days before his 40th birthday, swathed in a blanket of bone, Harry Eastlack ultimately died of respiratory failure resulting from complications of bronchial pneumonia.[18]

These kinds of tragic medical cases show just how unforgivingly deterministic genes can be. Harry Eastlack's misfortunes were the result of a rare condition called fibrodysplasia ossificans progressive (FOP), which tricked the body into growing bone wherever it was repairing some other tissue damage, such as from a cut or bruise. The ultimate cause of this condition was probably that Harry Eastlack was unfortunate enough to be born with a single adenine (A) nucleotide instead of a guanine (G) one at the precise wrong place—specifically, in codon 206, in the activin A receptor, type I gene (ACVR1), which is located on chromosome 2.[19] His horrible disease was thus caused by a single typo in his 6-billion-letter genome. If his genome was written down on paper, it would be the length of about 800 Bibles, and this single letter in the wrong place—what sounds like the most trivial of errors—rendered Eastlack a human statue, unable to move or speak, and ultimately, to even breathe.

It is difficult not to become fatalistic when we reflect upon cases like Eastlack's. We all are born with these kinds of *de novo* mutations—genetic typos that our parents did not possess themselves. On average, we each have several dozens of these, and the only thing that distinguishes the typos that we have from Eastlack's are where they lie along our genomes. Fortunately, for most people these typos have relatively little impact on their health. But there are many thousands of Mendelian disorders like FOP—that is, disorders that are caused by a single gene—and although each single disorder is typically very rare (FOP appears in only about one out of every 2,000,000 births), collectively these fatal little flaws constitute approximately 1 percent of all adult admissions to the hospital.[20] When we consider these cases, the moment of conception can feel very much like a game of Russian roulette in which we inherit a set of genes that may or may not ultimately kill us.

With these Mendelian disorders, there's a clear, direct, and rather deterministic way that the genotype (such as the single mutation in Eastlack's genome) produces its phenotype (his crippling disease). This kind of genetic influence is easy to understand and to study—the gene causes the disease—and it is frightening to consider its implications. I call the way that we think of this kind of strong genetic influence "switch-thinking": we think genes are like switches that control the phenotype. Eastlack's switch for FOP was in the "on" position, so he couldn't help but develop the disease. Most everyone else has the switch for FOP in the "off" position, so they never need to be concerned that their skeleton will fuse together as Eastlack's did. The many other Mendelian diseases work in pretty much the same way: if you have the genetic switch turned "on" for diseases like Huntington's disease and cystic fibrosis, you're almost certainly going to develop these, whereas if your genetic switch is off, you won't.[21] Switch-thinking is an extremely straightforward way to think about genetic cause and effect. But more often than not, it's also very misleading.

## Web-Thinking and the Evidence for Interactionism

In reality, there's actually a broad continuum of ways that genes influence us. On the terrifying end of the continuum is Harry Eastlack and what represents a case of strong genetic influence. On the other end of this continuum are cases of what Eric Turkheimer calls weak genetic influence.[22] On this end, the phenotype (such as whether or not one gets divorced) is associated with a genotype (identical twins do have more similar divorce rates than fraternal twins), but not at all in a direct, comprehensible, or deterministic way. In these cases, many genes (sometimes thousands) operate in concert, each contributing to the likelihood of the phenotype emerging. And all these genes only influence the phenotype when they actually create proteins, so the phenotype is also dependent on environment, experiences, and epigenetic influences from parents. This almost infinite number of interactions among this vast network of genes and experiences gives rise to complex human traits such as intelligence, risk of depression, shopping habits, and political attitudes.

Unlike Mendelian diseases, in these cases the underlying biological mechanisms are not yet understood, even by geneticists, and it's quite possible that they will never be understood in a tractable way. Genes are clearly implicated for complex human traits; these traits tend to be just as heritable as any other kind of trait. But the influence of any single gene is incredibly weak.

For the vast majority of human conditions, the weight of the accumulating evidence from genome-wide association studies forms such a consistent pattern that behavioral geneticists have posited what they call "the fourth law of behavioral genetics." This law states that "A typical human behavioral trait is associated with very many genetic variants, each of which accounts for a very small percentage of the behavioral variability."[23] The journalist David Dobbs prefers the much more pithy term MAGOTS—"many assorted genes of tiny significance."[24]

When considering weak genetic influences, we need to use what I call "web-thinking." We need to attend to the complex web of the interactions among genetic, epigenetic, and environmental influences that ultimately come to shape our phenotypes. Each of the many different influences on the web pulls us in a particular direction, but none so directly as switch-thinking suggests. So some aspects of your genome might pull you a bit in one direction, but the fact that your mother breastfed you as a child pulls you in another. Your relation with your best friend tugs you in yet another direction, as does your newly acquired penchant for French cheeses. In the end, you are a product of all of the different influences on your life. With web-thinking, phenotypes are not predetermined; it is possible to change your position on the web, to move across it in different directions, as you develop. The web might be tilted so that it's easier to move in one direction than the others, but ultimately there is no single force that is determining your position. The web offers resistance, but it still allows freedom.

## The Case of Human Height

An example of a trait which we should consider with web-thinking is height. What causes someone to be as tall as he or she is? Consider the case of Muggsy Bogues, the shortest person to ever play in the NBA. If he stretched himself out as tall as he could, he would just reach 5' 3". In 1988, he played for the Washington Bullets, where he was frequently paired up with his teammate, Manute Bol, who, at 7' 7", was the second-tallest person to ever play in the NBA. So why was Manute Bol so much taller than Muggsy Bogues? Well, that's easy—just look at their parents. Bogues's father was 5' 6" tall, and his mother was only 4' 11". In contrast, Bol's father looked down from a height of 6' 8", and his mother was a towering 6' 10". Clearly, Bol inherited genes for much greater tallness from his parents than did Bogues, and that's the best explanation for their striking difference in height. As discussed above, height is highly heritable, with estimates in the 80 percent to 90 percent range.[25] This means that if the difference between the tallest and shortest 5 percent of the population is

about 11 inches, then about 10 inches of this gap should be accounted
for by genetic factors, with the remaining 1 inch accounted for by the
environment.[26]

Yet explaining why Bol is so much taller than Bogues becomes
considerably more difficult when we try to specify *which* precise genes
are responsible for the difference. Genes strongly influence height,
but no individual gene contributes much to our height at all—the
single gene that is most strongly associated with height only adds
about one-sixth of an inch to the height of those who possess it.[27] A
recent genome-wide association study into height investigated all of
the SNPs that were related to people's height. The researchers found
294,831! This is a depressingly large number of genetic variants—
in fact, it's more than 12 times the number of genes that we have,
indicating that, as geneticist David Goldstein puts it, "most genes
are height genes."[28] Worse than that, all of these many thousands of
variants combined explain less than half of the variability in human
height.[29] This means that if you wanted to genetically engineer a
designer baby who you would like to grow up to be tall, you would
have to make almost 300,000 genetic alterations to the embryo, and
you still would only be halfway there. When the genetic evidence
suggests that almost all genes are relevant to height, then in a way,
we learn close to nothing about the genetic basis of height. The irony
is that we can see the genetic contribution to height far better by
simply attending a Bol–Bogues family reunion than by using the
most cutting-edge technology to look at their actual genomes.

One reason for the difficulty here is that height is a developmental trait that interacts with the environment as you develop. Your
adult height is not directly transcribed into your genome; rather,
your genome shapes how you will respond to environmental events
that influence your height over the course of your development.
This means that we actually overestimate the genetic contribution of
height because twin and adoptee studies make the assumption that
genetic and environmental influences are separate; but because genes

and environments can interact, and these interactions get lumped in together with the genetic estimates, the genetic estimates are exaggerated.[30] Moreover, because heritability estimates are almost always based on people living in one culture, who share much of the same environment, they don't speak to any differences in environments between cultures, which can sometimes be extremely pronounced. Heritability estimates thus entirely leave out cultural factors that may contribute to a phenotype, which also leads to estimates that overstate the amount of a genetic contribution to a trait. This can clearly be seen when we move from comparing the height of two individuals living within the same culture, like Bogues and Bol, to the average height of people living in two cultures, like the Netherlands and Japan.

After I received my bachelor's degree, I had the wonderful experience of becoming a junior high school English teacher for a couple of years in a small town in Japan's Nagasaki prefecture called, of all things, Obama. One of the many highlights of this experience for me was that this was the first time in my life that I got to feel tall. I'm 5' 8", which is about 2 inches shorter than the North American average for men, so I'm used to being one of the shorter men in a group. Well, I was one of the tallest people in the town of Obama, and the whole place seemed to be built on a smaller scale. I would regularly bump my head on the top of the doorframe of my apartment, children would ask me if I could show them how to slam-dunk a basketball, and I was told that I couldn't buy shoes at the shoe store in town because I was simply too big. I loved it! Contrast this with my experiences in the Netherlands, where I've collaborated with several psychologists. The Dutch are the tallest people in the world. This became painfully evident to me as the first Dutch student that I worked with, a woman, was about 4 inches taller than me. The average height today of Dutch men is 6' 1", about 3 inches taller than the average American man. Because height is so clearly influenced by genes, it follows that Dutch people should have many more "height genes" than Americans. But the story behind national differences in height is considerably more complex, and searching for the elusive Dutch height genes doesn't get us very far.[31]

First, we need to consider whether the Dutch-American difference in height has been stable across time. It hasn't. In 1865, the average American man stood at 5' 8", which was about 3 inches *taller* than the average Dutch man at the time, who stood at only 5' 5".[32] So if geneticists wanted to understand the reasons for this height difference in 1865, they may well have assumed that Americans had more height genes than the Dutch. Instead, it seems that lifestyle changes are responsible for the recent Dutch growth spurt. In the late 19th century, Americans had the third highest GNP per capita, and they were among the tallest people in the world. During the same period, the Netherlands was going through an extended economic downturn, and as the country recovered, the average height of the population grew accordingly. With wealth comes a better diet, and during wealthier periods, people are more likely to get the necessary nutrition during critical growth spurts in infancy and in adolescence. (One striking example of the tight link between wealth and height can be seen in that Japanese children who were raised in California in 1957 were about 5 inches taller than their compatriots residing in Japan).[33] Curiously, while the rest of the world has largely been getting taller alongside economic development, the average height of Americans has stagnated over the past few decades, despite a strong growth in average annual income over this period. One theory for this is that the large amount of fast-food consumption in the United States has resulted in the growth axis of Americans being shifted from vertical growth to horizontal growth in the waistline.[34]

One aspect of diet in particular is associated with height: dairy consumption. Research with Americans finds that those who report drinking a lot of milk as a child and as a teenager are approximately half an inch taller than those who don't recall drinking so much, controlling for other variables such as race, income, and education.[35]

The Dutch have one of the highest rates of milk consumption in the world, and in general there's a strong positive correlation between annual milk consumption and the average height of nations.[36] Over the past few decades, Japan has gone through a national growth spurt, gaining close to 4 inches in height.[37] To my chagrin, this has pretty much eliminated the height advantage I once enjoyed in Japan. Much of this increase in height has been explained by the increased consumption of milk as part of the National School Lunch Program,[38] which first began to be implemented in 1956, although Japanese milk consumption still remains at less than one-fourth that of Dutch levels.

Moreover, the effect of milk consumption on height varies with the seasons; humans, just like plants, grow more in the spring and summer than in the fall and winter. This seasonal variation appears to be due to ultraviolet radiation from the sun that synthesizes vitamin $D_3$ from cow's milk (and is more concentrated in cow's milk in the summer months). All of these factors contribute to the speed of our growth.[39]

This means that population differences in height cannot simply be explained by population differences in genes. Cultural differences in average height are substantial, and they overwhelm the estimates for environmental contributions to height from twin and adoption studies that are conducted within a single culture. Remember, within a culture, environment only accounts for about 1 inch of height between the tallest and shortest 5th percentiles in a distribution. Yet Japanese moving to America in the 1950s got a 5-inch boost in their height. This highlights an important, yet counterintuitive point: even if genes cause much of the individual difference in a trait such as height within one population, this does not mean that any group differences between populations are due to those same genes. For example, skin color varies around the world, with darker shades— and the genes associated with darker skin—being more common in areas closer to the equator.[40] Yet if you were to compare the skin color of European-descent people living in cloudy Seattle with that of European-descent people living in sunny Los Angeles, you would likely find that the Seattleites are paler than the Angelenos. This population difference would emerge even if the ancestral population of these two groups was from identical genetic stock. The difference in average skin colors between Seattleites and Angelenos is due to the different frequencies of sunny days, yet the difference in skin colors among people living in Seattle would largely be due to genetic differences. Even though genes lead to individual variability in skin color, it does not follow that population variability in skin color is solely due to the same genetic factors.

Height is thus a prototypic case of weak genetic influence, and the only proper way to understand it is through web-thinking. Although genes are strongly associated with height, they are not at all like switches that specify some units of growth. Rather, your height is a result of an ongoing interaction of many, many genes being influenced by various aspects of your environment, which affects their expression throughout your development. The complexity of the genetics of height confounds any simple switch-thinking account of height genes.

## Cognitive Misers and the Appeal of Switch-Thinking

But of course, trying to understand a system as vastly complicated as the genetics of height is quite overwhelming and rather discouraging. And this is where we run into trouble. There have been many psychology experiments that show that people can be exemplary thinkers when they are highly motivated and feel that they have the ability to understand things. In these situations, people typically will buckle down, consider all the information that they encounter until they understand it fully, and then make a carefully reasoned judgment.

But when people aren't motivated, such as when they are considering a new proposed law that won't affect them directly, or when they feel that they don't have the ability to fully understand a complex problem, then they show a curious response: they start to attend to things that are really quite irrelevant to the quality of the information. They'll be persuaded just because an argument comes from an attractive person, or from someone who has an impressive sounding title, or by whatever information is presented first, or by a snappy sound-bite that sticks in their mind. That is, when a problem is too challenging, we try to conserve our cognitive resources to make a judgment that feels "good enough."[41] We become what psychologists call a "cognitive miser." Thinking takes effort, especially for complex problems, so we use mental shortcuts—what psychologists term "heuristics"—that appears to give us a satisfying answer.

When estimating the frequency of an event, for example, we often base our estimates on how many relevant examples we can bring to mind. This is known as the "availability heuristic." If we want to estimate the likelihood of being killed by a shark versus being killed by a cow, we think of the number of incidents of each of these that we know. Because fatal shark attacks are so noteworthy, they receive much more press attention than fatal encounters with cows, and thus most people can more readily bring to mind examples of shark-caused deaths than cow ones. Our estimates of the ratio of shark and cow attacks are also influenced by how well an event fits with our existing beliefs—this is known as the "representativeness heuristic." Some ideas just are easier to imagine because we have some preexisting expectations and stereotypes that guide our thinking. When it comes to sharks, we think of them as all teeth—efficient cold-blooded killing machines—and we can easily imagine how they could tear up a lone swimmer or surfer. In contrast, people usually think of cows as lumbering docile grazers—gentle beasts that seem content chewing their cud. Attacking humans just doesn't seem to be *representative* cow behavior. The availability and the representativeness heuristics make us feel that shark-caused deaths are more common than cow-caused deaths, when in actuality, deaths from cows are far more common. Our mental heuristics provide us with quick, satisfying, and often completely incorrect answers.

We're cognitive misers when we approach challenging problems, and the ways that genes influence complex traits like height is about as impenetrable as a question can get. So, in the face of such an intractably complex problem, we respond to it by instead replacing it with a much simpler imagined one. We shift from web-thinking to switch-thinking. Rather than conceiving of height as a complex web of interacting influences, it's much more straightforward to just think that tall people, like Manute Bol and the citizens of the Netherlands, have "height genes," or even "*the* height gene," and short people do not. By making this swap, we're left with the satisfying feeling that we do in fact understand how genes affect us. We've changed the problem to one that makes intuitive sense to us.

Switch-thinking is a reasonably accurate way for making sense of monogenic (that is, caused by a single gene) diseases like FOP, but its utility is extremely limited. The tangled web of interactions between our genes and our environment that influences our height is far more common when it comes to genetic influences. Only about 2 percent of genetic diseases are caused by a single gene;[42] the norm is that diseases are massively polygenic. And the same is true for the vast majority of human characteristics. For example, the world's largest consumer genomics company, 23andMe, which has genotyped more than one million people, provides genetic information on about 60 human traits, such as whether your hair is curly, whether you can taste bitter foods, whether you can smell asparagus in your urine, have fast-twitch muscle fibers, have wet earwax, or are lactose intolerant. Of those 60 traits, only one of these, whether you have wet or dry earwax, is a pure Mendelian trait. Simple genetic explanations are the exception, not the rule, and we are making an inductive error when we generalize from them.

Yet switch-thinking explanations are what students primarily learn in high school and even university courses on genetics, and this is how genetic discoveries are usually described in the media. In high school you might have been taught, as I was, a switch-thinking story about the simple Mendelian basis of eye color, or of whether you could roll your tongue into a tube. Yet even these two classic examples don't hold up to closer scrutiny—eye color is determined by several genes, and the geneticist who originally identified tongue-rolling as a Mendelian trait backtracked from this claim 50 years ago, and added that he was "embarrassed to see it listed in some current works as an established Mendelian case."[43] But high school genetic classes have relied upon these false examples because there really aren't any obvious Mendelian traits for students to measure.

Gregor Mendel's discoveries of how characteristics of peas were inherited is a prototypic example of switch-thinking, and although it's accurate for understanding his experiments, it's worth noting just how lucky Mendel was in finding the straightforward patterns that he did. In his experiments, Mendel recorded the proliferation of seven different pea traits. Each of these traits was determined by a single gene, they were each located on a different chromosome (with the exception of one pair that was located on opposite ends of a single chromosome) so they were inherited independently from each other, the traits didn't blend, and the different traits didn't interact in any way with each other. There are actually many other characteristics of peas that don't operate in direct, switch-thinking ways. Indeed, Mendel initially had identified fifteen traits for study, but over time reduced this list to seven, likely because it was only for these seven traits that he obtained an interpretable pattern.[44] He had also continued his breeding efforts with several other species, such as hawkweed, different species of beans, thistles, toadflax, and bees, but rarely was he able to replicate his clear findings with peas. Indeed, later in his life, in the face of a much messier pattern of data across his different breeding studies, Mendel seemed to lose faith in his original findings.[45]

Like the proverbial drunk who only looks for his keys under the streetlamp because that's where the light is good, we make sense of genetic accounts through switch-thinking explanations because we can easily understand them. In the vast majority of cases, however, switch-thinking is just plain wrong. Yet we continue to rely on switch-thinking, not only because it's a simpler way of thinking about a complicated problem, but because it also resonates with a deep human impulse for how we make sense of ourselves and our worlds. As the next chapter discusses, switch-thinking gets at the very essence of our essences.

# 3.

# Your Genes, Your Soul?

IN GREECE IN THE FIFTH CENTURY BCE, HIP-
pocrates laid the foundational theory of Western
medicine that would stand for more than 2,000
years. He traced the source of all diseases to the
balance of four fluids that coursed through our
bodies: phlegm, blood, yellow bile, and black bile. Too much, or
too little, of any of these humors, and the body would fall out of
homeostasis and become ill. By offering this framework, Hippocrates rejected earlier views that disease was caused by superstition
or gods and instead provided a natural basis to understand it. The
far-reaching effects of these four humors were elaborated upon a
few centuries later by the Roman doctor Galen, who maintained
that people's personality traits were also a function of the balance
of these four fluids. People who were extroverted and outgoing were
so because they had a surplus of blood, whereas an excess amount
of black bile led one to be despondent and analytical. This was the

leading psychological framework throughout most of Western history, and although it has largely fallen from use in the past century (indeed, there's no supportive evidence between these fluids and personality traits), its popularity has been revived in a few books by Tim LaHaye, the Evangelical Christian minister and author of a popular series of books about the upcoming Rapture.

Hippocrates's and Galen's influential ideas revealed a quest to understand our health and our personalities by looking for an underlying *essence*—some kind of inner force that determines our nature. Similar efforts to identify the underlying essence of character were commonplace in the 19th century, when a variety of disciplines emerged that promised to divine your underlying temperament from assorted bodily features. Phrenologists would survey the bumps and dents along the topography of your head, based on a belief that any protuberances indicated that one of 30 underlying brain "organs," such as organs of destructiveness, benevolence, or mirthfulness, guided your character. Likewise, craniometrists used a series of sophisticated calipers to measure the overall shape and capacity of the skull to discern your underlying traits and intellectual capacities, whereas physiognomists would determine your true character by studying the shape of your facial features. For example, the 19th-century Italian criminologist Cesare Lombroso described how he could identify criminals by their faces alone: "Thieves are notable for their expressive faces . . . , small wandering eyes that are often oblique in form, thick and close eyebrows, distorted or squashed noses, thin beards and hair, and sloping foreheads. Like rapists, they often have jug ears."[1] By the early 20th century, some psychologists predicted people's temperaments on the basis of their somatotypes—that is, their body shape—with thin ectomorphs thought to be intelligent, fearful, and asocial, wide-set endomorphs expected to be sloppy, sociable, and lazy, and muscular mesomorphs perceived as ambitious and hardworking. Similarly, folk psychology in Japan has it that your blood type determines your personality and dictates the kinds of people with whom you can get along. Somehow the proteins that cling to the surface of red blood cells are thought to make a person who he or she is. My blood type is type O, which means that I'm supposed to be optimistic, strong willed, sociable, self-centered, flighty, a workaholic, and that I particularly get along with people who are either type AB or O. When I tell Japanese people that my blood is type O, they often respond with a knowing nod and say, "I thought so."

These paradigms, from Hippocrates's humors to Japanese blood types, have ultimately all failed at accurately predicting our health and personalities for want of supporting evidence. But they share the belief that some kind of biological essence makes us who we are, without ever specifying precisely the mechanisms through which these biological essences get translated into our psychological characteristics.

We still are searching for these magical hidden and underlying essences. But now, in the 21st century, most scientists believe they've found that mysterious essence: it's our genes. As we'll see, it's our attraction to the idea of essences that makes genetic thinking so problematic. But to understand how people think about genes we first need to understand how people think about essences.

## Thinking about Essences

Why did the Dow Jones drop 50 points yesterday? Why was Typhoon Haiyan so destructive? And why did Donald Trump win the presidency? The world unfolds around us, and we try our best to make sense of it. The psycholinguist John Macnamara declared that "the human mind is essentially an explanation seeker."[2] In many ways, our minds guide us to think like scientists—always looking for underlying causes for past events so that we can better predict and control future events. And the underlying causes that we find

most attractive—no matter how complex the web of factors involved
in any event—are ones that posit a fundamental, internal, and
hidden essence.

Going back to the beginning of Western philosophy, scholars
have sought the supposed essences that underlie the nature of reality. What is it that makes a leopard a leopard, and what makes it
different from a lion? Aristotle famously proposed that each entity
possesses an essence that ultimately makes it what it is, and that
without such an essence, the entity would no longer be itself. John
Locke describes an essence as "the very being of anything, whereby it
is what it is."[3] Essences form the basis of identity for any entity, and
they are the *ultimate cause* for why those entities function as they do.
So the essence of a dog gives rise to its furry frame, love for bones,
keen sense of smell, unfaltering loyalty, and desire to play fetch. If
a dog's essence could somehow be surgically removed, it would lose
these attributes, and would no longer seem like a dog as we know it.
This idea has been enormously influential throughout philosophy,
although it's hotly contested whether such essences actually exist in
any meaningful way.

Meanwhile, psychologists have sidestepped this thorny philosophical debate by noting that regardless of what the world is actually like,
people act as though they *believe* it is made up of underlying essences.[4]
We find the notion that invisible internal forces make things as they
are to be quite irresistible. Take the enduring appeal of the story of
Harry Potter. Harry's biological parents were wizards, but when they
were killed by Voldemort when Harry was just a baby, he was adopted
and raised by his Muggle (nonwizard) uncle and aunt. Harry's uncle
and aunt were determined to never let Harry know anything about
the magic that his parents possessed. But despite their efforts, Harry
couldn't keep his own magical essence repressed. Without even realizing it, he magically caused things to happen around him: for example, he made the cage around a snake disappear. Even though he
was raised as a Muggle and had no knowledge that his parents were

wizards or that wizards even existed, Harry could never be a Muggle himself. His "wizard" essence could not be suppressed by whatever beliefs he had or events he experienced. It ultimately was his nature. And this notion that Harry would become who he was because of some inherent kind of "wizard" essence is highly intuitive to us. We grasp it immediately without much explanation because we believe that essences make up the world around us.

This belief that things are as they are because of their underlying essences is called *essentialism*, and it is one of the most persistent and widely documented psychological biases.[5] *Essentialism* has a few notable characteristics that, as we'll see, ultimately guide how we think about genes. First, people tend to think of essences as being *deep down and internal*; they're typically not visible at the surface, but lie invisibly at the core of an entity. So if we think that a leopard has spots because of its underlying leopard essence, then even if we were able to remove the spots via dye or surgery, we wouldn't expect this to impact the leopard's essence itself; the animal would still be a leopard, just one without spots. We think of the essence as causing the spots, but it does not exist in the spots themselves. In our minds, essences remain deep and buried. One consequence of this is that we view changes to the inside of an organism as more impactful than changes to the outside.

This tendency is the source of our enduring belief that "You are what you eat." Of course, when people say this, they don't literally believe that they are living in a body that is made up of, say, pizza and beer. But they do seem to believe that ingesting the "essence" of different foods can affect their own identity. Take the following study. American students were asked some questions about a mythical island tribe called the Chandorans.[6] The students learned that the Chandoran men hunted only two kinds of animals: sea turtles and wild boars. However, the students read one of two different sets of facts about the Chandoran diet. One group read that the Chandorans ate the meat from the sea turtles that they hunted, but they didn't eat any of the meat from the boars; the boars were hunted only for their tusks, which were used to make tools and decorations. The other group read the opposite—the Chandorans ate the meat from the boars that they hunted, but they didn't eat the turtle meat; the turtles were only hunted for their shells. After reading one of these two possible descriptions, the students were asked to describe some characteristics of the Chandorans. The turtle-eating Chandorans were described as being better swimmers and less aggressive than the boar-eating ones, even though both descriptions of the Chandorans were identical aside from what animals they were described as eating. Regardless of the descriptions that people read, the turtle-eating Chandorans were suspected to be more turtle-like, and the boar-eating ones more boar-like. This notion that we can acquire the essences of the foods we eat is widespread across cultures. For example, traditional Chinese medicine has it that eating a tiger penis will enhance your virility, and several tribes from around the world have practiced various forms of cannibalism with the belief that this allows them to acquire the characteristics of the deceased.[7] Essences can be most effectively acquired when we ingest them and they become part of our own insides, where they remain deep and hidden.

Another feature of essences is that we think of them as *natural*; they are somehow just meant to be. This means we are more likely to turn to essences to understand entities that are *natural kinds*— that is, things that are observable in nature. We are much less likely to believe in essences underlying artifacts created by humans. For example, we are more committed to the idea that natural categories like "rabbits" or "diamonds" have a fundamental underlying essence, than we are to the idea that nonnatural categories like "shirts" have an underlying essence. Of course, we very well might be hard-pressed to describe precisely what are the essences that underlie rabbits or diamonds. But even without being able to see them or describe them, we still tend to think that such essences really do exist. An understanding about the nature of these essences seems potentially knowable, and we believe that appropriately trained experts would be able to identify them. So, if we saw an animal in a zoo that looked like an unfamiliar kind of rabbit, but an expert told us that it actually wasn't a rabbit at all, but was a species of miniature kangaroo, we would probably trust the expert. Or if an expert told us that a sparkly gem wasn't really a diamond, but was actually a cubic zirconium, we would likely believe such a distinction. We would trust the experts because we'd believe that they have the knowledge about rabbit and diamond essences that we do not. In contrast, this same kind of expertise does not have the same power for human-made artifacts. If an expert told you that what you were wearing was not really a "shirt" but was actually a "blouse," you very well might not have the same faith in that distinction. This is because we expect that the categories of natural kinds, like rabbits and diamonds, are based on their underlying essences, and these are what the experts are identifying. Shirts and blouses, in contrast, differ from each other because of conventions, which often change over time and can be debated among experts, rather than through some hidden underlying force.[8]

The perceived naturalness of essences leads us to feel that it would be wrong to tamper with them. If there were a way to change the essence of something, such as casting an alchemist's spell to turn cubic zirconia into diamonds, or by performing whatever necessary microsurgeries to change a kangaroo's essence into that of a rabbit, it likely would strike you as being deeply unnatural. This reaction against artificial modifications to supposed essences only intensifies when it comes to our genes, as we'll see shortly.

A third feature of essences is that we use them to draw the *boundaries of categories*, such as when we judge whether an animal is a kind of kangaroo or rabbit. We assume that members of a category share a common essence, whereas members of different categories do not. If you select any two rabbits at random, they will share a common rabbit essence, and that essence will be fundamentally different from other kinds of animals—even of miniature kangaroos that look like rabbits. And because we think that members of the same category share the same essence, we are likely to generalize from one instance of the category to another. So if we learn, say, that a particular rabbit is nocturnal and eats tree bark, then we assume that other rabbits of that same species would also be nocturnal and would eat tree bark. We believe that the essence they share gives rise to their shared behaviors. In contrast, knowing that a particular rabbit is nocturnal and eats bark would not lead us to assume anything at all about the behavior of a miniature kangaroo, because we assume they have different essences. These perceived boundaries of essences are what allow nature to be carved at its joints. Indeed, the taxonomic system developed by the naturalist Carl Linnaeus, which is still used today to identify all the kinds of species, was built on the idea that species differed from each other in terms of their underlying essences.[9]

A last notable feature about essences is that we tend to think of them as largely *immutable*. So we might think that the essence of a leopard includes a ferocious temperament, but even if a leopard was raised by loving humans as a pet, and never acted in a ferocious way, most people would probably suspect that its underlying leopard essence had remained intact. And if we heard that the pet leopard one day unexpectedly attacked the mailman, this likely wouldn't strike us as all that surprising; it was just a matter of its immutable leopard essence breaking through, much like Harry Potter's magical abilities appearing despite his Muggle upbringing. Whatever socialization may do to help control the expression of the underlying essence, the essence itself always remains untouchable.

Claire Sylvia's life quite dramatically reflects this immutable aspect of essences. A former professional dancer who was always very health-conscious about her diet, she developed primary pulmonary hypertension, and at the age of 47 she was the first recipient of a heart-lung transplant in New England. She received the organs from an 18-year-old boy who had been killed in a motorcycle accident. At the time of his death, the boy had a half-eaten package of chicken nuggets in his pockets, and he was known to have a fondness for beer. Although Claire had liked neither of these before, after her transplant she developed profound cravings for chicken nuggets and beer. Claire reported that she had started to feel more masculine and aggressive than before, and her daughter asked her why her walk had changed such that she was now "lumbering—like a musclebound football player."[10] In her book, *A Change of Heart*, Claire writes about how she felt that she had acquired the essence of her donor—she describes a vivid dream where she kissed and then inhaled her donor.[11] Indeed, it's not uncommon for recipients of organ transplants to report feeling that they have taken on the essence of their donors, together with assorted characteristics—one study found that a full third of heart transplant recipients felt they had taken on some aspects of their donor's personalities.[12] Moreover, many people are wary about receiving heart transplants from morally suspect individuals for fear of becoming corrupted by their donor's immoral essences.[13] Sometimes this new imagined essence can seem to take over people's lives: in one extreme case, after a Georgian man named Sonny Graham had received a heart transplant, he ended up not only marrying the donor's wife, but he ultimately committed suicide by shooting himself in the throat with a shotgun—the same way that his organ donor had earlier killed himself.[14] Somehow, it seemed that the course of Sonny's life was inescapably tied to that of the original owner of his new heart.

Now, I'm not suggesting that organ recipients really are inheriting the original essences of their donors, but some of them clearly believe that they are, and they're acting accordingly. So strong is the sense of the immutability of essences that it's easy to imagine that these donor's hearts will forever be imbued with the fundamental qualities of the donors, such as their food preferences or suicidal fates, regardless of the characteristics of their new owners. People view essences to be immutable enough that they can even appear to permanently adhere to objects that are transferred to someone else. For example,

imagine that you're a participant in the following study: you're asked
whether you'd be willing to try on a sweater that used to be worn by
Hitler. Would you be willing to do it? Would it make a difference
to you if the sweater had been thoroughly dry-cleaned? Many of the
students in the actual study showed a clear aversion to the idea of
trying on Hitler's sweater, and no amount of cleaning was thought to
be able to fully bleach out its evil essence—they balked at the idea of
pulling the seemingly contaminated sweater over their faces.[15] Somehow, Hitler's essence seemed to have become forever woven into the
sweater's fabric. It's this same belief in the immutability of essences
that can help explain why people are so attracted to things that were
once owned by celebrities. Although it may seem insane for someone to pay $48,875 for a tape measure just because it was previously
owned by John F. Kennedy, as a Manhattan interior designer did, it
makes more sense when we consider the purchase from the perspective of essentialism. This is no longer just an ordinary tape measure;
it is one that has forever been transformed by the indelible essence
of Camelot. And the new owner now possesses some of that essence.

These are the features that make up essentialism—essences are
*deep down and internal, natural, immutable, ultimate causes,* and they
*carve the boundaries* of the natural world. And essentialism appears to
be a universal human trait. In every one of the many cultural groups
in which essentialism has been investigated, there is clear evidence
for it.[16] For example, the anthropologist Francisco Gil-White spent
time in the Mongolian steppes where he lived among the two ethnic
groups there, the Torguuds and the Uryankhais. Although these two
groups look highly similar, and have quite comparable lifestyles, the
Uryankhais are believed to be able to cast curses, whereas Torguuds
cannot. Gil-White posed a Harry Potter–type question to one of his
informants and asked what would happen if a boy born of curse-cast-
ing Uryankhai parents was raised by nonmagical Torguuds. Just as in
the Harry Potter series that would later be published, his informant
replied that, by dint of his birth to Uryankhais, the boy would have

the capability to cast curses, yet because he was raised as a Torguud, he would never be aware of his talents. His Uryankhai essence couldn't be denied by either his knowledge that he was a Torguud, or by his exclusively Torguud experiences—it would inevitably shape who he would become.[17] This kind of evidence for essence-based thinking is found across the world, although some kinds of people are more committed to thinking about essences than others.[18] For example, wealthy people show more evidence for essence-based thinking than poor people; those with a lot of money tend to find the idea that certain people succeed because of innate talent, rather than the circumstances they're born into, to be very compelling. In contrast, poorer people remain all too aware of the environmental challenges that they face.[19] Likewise, in India, high-caste Indians view the characteristics associated with caste to be innate, whereas those from lower castes view the characteristics to be a product of socialization.[20]

Further evidence for the universality of essence-based thinking is found in very young children. At the age when children are just starting to offer explanations about their world, they already show a preference for seeing essences in natural things but not in human artifacts. For example, when kindergarten students were told that a doctor had taken a raccoon, shaved away some fur, dyed it black with a white stripe down the center, and then surgically installed a sac of stinky fluid, most children insisted that the animal was still a raccoon and not a skunk; its raccoon essence was thought to remain largely intact beneath its skunk dressing. In contrast, when the same children were told about how doctors took a coffeepot, sawed off some parts, and added some others to make it look and function just like a birdfeeder, the children were now quite comfortable with declaring that the coffeepot has been transformed into a birdfeeder.[21] The children recognized that the coffeepot, as a human artifact, didn't possess an essence that could withstand the transformations. Likewise, young children also share the sense that objects can acquire their owner's essences: for example, one study found that preschool children viewed a flag pin to be far more valuable if it had earlier been worn by President Obama.[22] It's entirely intuitive to children that there's something about the president that becomes a permanent part of a pin that he's worn.

That essence-based thinking is prevalent across all cultures that it has been studied in, and that it shows up so early in life, suggests that essentialism itself is, well, part of *our* essence. Some researchers have argued that we evolved to be essence-based thinkers—humans have spent most of their ancestral past as hunters and gatherers, and it would have been adaptive for our ancestors to categorize the natural world around them into different species, based on the members of a species sharing a distinctive set of characteristics. Being able to identify that, say, a particular bird was a ptarmigan and not a grouse, would enable hunters to better predict its behaviors, which would have enhanced their hunting prowess.[23] We then generalized this essential way of thinking to other natural substances (like diamonds) and to ethnic groups (like the Uryankhai). Regardless of the reasons for how humans became so adamantly essence-based in their thinking, the evidence is clear that an attraction to essences as explanations is a very widespread and ingrained tendency.

## Essences and Placeholders

But what could these essences that guide our reasoning actually be? What precisely *is* the essence that is woven into Hitler's sweater, pulses in Claire Sylvia's heart, allows Uryankhais to cast curses, fans the aggression of the boar-eating Chandorans, or distinguishes diamonds from cubic zirconia? These essences certainly are beyond our own abilities to observe or describe, yet they nonetheless have a powerful psychological impact on us. The psychologist Doug Medin argues that it doesn't really matter that we can't conceptualize what these essences are actually like because we apply a psychological shortcut to get around the problem. In our minds, when we think about essences, we replace the essence with an *essence placeholder*. By using a little psychological "bait and switch," we swap the notion of an abstract and elusive essence with a placeholder that seems more concrete and potentially scientifically justifiable. These placeholders are perceived to share all of the properties of essences themselves. And we then turn to these essence placeholders whenever we want to explain why things are as they are.[24]

So what do people use for essence placeholders? Hippocrates used the four humors as essence placeholders. The underlying essence that caused a person's disease was, say, an overabundance of black bile; a concrete substance that sounds like it could plausibly lead to the manifest disease, yet retains the powerful and mysterious properties of an essence. Likewise, phrenologists replaced essences with the placeholders of imagined underlying brain organs, such as mirthfulness and destructiveness; and the placeholders for folk Japanese beliefs about personality are people's blood types. You can find these kinds of essence placeholders throughout the world, from Polynesian accounts of the all-pervasive power of *mana*, Western African beliefs in magical *juju*, to the Judeo-Christian concept of the "soul." But it is hard to imagine a more fitting essence placeholder than the concept of the gene.

Genes, or at least the way that most people think about genes, match all of the key attributes of essences. They remain invisible, deep down and internal, yet we know that they're there. Our bodies are not actually made up of genes per se—they are constructed out of proteins, and genes are the hidden force that ultimately guides the creation of those proteins. Even though we never see genes directly ourselves, we trust that genes can be identified by geneticists who possess the expertise and technical equipment that we don't have. People understand that genes, like essences, are the ultimate causes; they are a key reason behind why someone looks and acts as he or she does.

Moreover, as with essences, people think of genes as natural; they are only relevant for natural kinds, such as species. Genes have no place in human artifacts, which exist because someone has assembled together a set of parts. It is telling to look at the way that people typically respond to the intersection of human artifacts and genes, such as when genetically modified organisms are created in a lab. Quite often, people are deeply bothered by the idea of genes being artificially manipulated. These creations can feel intensely *un*natural and disturbing. People seem to only find natural those genetic combinations that have emerged without human intervention. Genes are also perceived as the means by which biological categories exist—they are seen to carve nature at its joints. Rabbits are understood to have rabbit genes, which are thought to be possessed by all other rabbits as well. Likewise, rabbit genes differ in some specific ways from kangaroo genes. Genes are thus seen as shared by those from the same species (or families or ethnic groups), and different from those of other groups. Finally, just as essences can be passed from wizards to Harry Potter, from John F. Kennedy to his tape measure, from boars to the Chandorans who eat them, and from heart donors to recipients, genes too are understood to be transmitted across generations; they are *immutable*.

Given that genes are thought of as the essences that undergird our own lives, they can often seem like the biological equivalent of the soul. Our genes are incredibly personal—barring identical twins, no one else has the exact same set of genes as you do. Like "the social security number that God gave you," as Stephen Colbert put it, genes register our one-off individuality. And we carry this badge of uniqueness from the time of conception. The very first instant that our existence had any kind of beginning, at the moment that an egg was fertilized, our genes were there in full; the complete set was laid out in the same fixed sequence that would remain for the rest of our lives. And as episodes of *CSI* remind us, even after death, genes remain as permanent identification cards; the tiniest speck of a person can speak from the grave and announce the owner's identity, long after other distinguishing markers have eroded away. And as with our souls, we have no choice in our genes, but they still seem to be the ultimate cause for how our lives will unfold. As the codiscoverer of DNA James Watson put it, "We used to think our fate was in the stars. Now we know, in large measure, our fate is in our genes."[25] The events in our lives appear to spring directly and inevitably from this mortal coil. There's no question genes are remarkably fitting essence placeholders.

But there is one key way that genes are unlike any other essence placeholders: they can be sequenced. So when direct-to-consumer genomics companies sell you the opportunity to peer into the arrangement of nucleotides that make up your genome, they seem to be offering you the chance to learn who you truly are. This is perhaps why the names of these companies so often include the word "me," as in "23andMe," "Knome" (often pronounced "Know me"), and "deCODEme." And because we often first encounter information about our own genes when we learn of a medical problem with them, this information can sometimes be deeply alienating. As the founder of the first genetic counseling master's program, Melissa Richter, put it, "When a genetic problem hits, it hits at the very gut of people, at the questions of what I am and what do I leave the world."[26]

Despite all the power we ascribe to our genes, we actually have a remarkably poor understanding of the science of genes. For example, consider this straightforward question: "Where do you think genes might be located in someone's body?" When this question was asked to a sample of American adults, a whole host of different responses were received. Less than half of those asked knew the correct answer, which is that our genes reside in our cells. The majority of those sampled either said they didn't know, or they gave an answer that was patently incorrect. The most common wrong answer that people offered was that genes are located in our brains—an answer that indicates that people see genes as affecting their thoughts and behaviors. Another common answer that people gave was to say that genes are "in the vagina area of the woman and penis area of the man"[27]—an answer that points to the heritability aspect of genes, revealing the understanding that genes are passed on to offspring.

What's curious about this is that, despite not really knowing much at all about what genes are, people talk about genetics a great deal, and they routinely refer to genes when trying to explain their social worlds. A lack of genetic literacy has not prevented the advertising industry from talking about a company's "DNA," nor has it discouraged people from casually talking about having things like "green thumb genes," or for praising the great genes of professional athletes. Research shows that even young children will make genetic attributions for people's behaviors well before they have received any kind of formal genetics education. For example, researchers in one study told children a story in which a baby born to "not so smart" parents was accidentally switched at birth and raised by "smart" parents. When asked about the switched child's future potential, children as young as fourth and fifth grade acknowledged that the child's birth parents would influence the child's intelligence. As one child said of the switched-at-birth child, "It will have trouble. It's in its genes."[28] People don't need much knowledge of how genes actually operate in order to use them to explain the world around them.

The reason that people can routinely make such kinds of genetic attributions in the absence of a real understanding of the mechanisms of genes is that usually when most people are talking about genes, they're *not* really talking about genes. They're talking about essences. And by swapping genes for essences, people come to think about genes in the same terms that they think about essences—that, somehow mysteriously and ineluctably, genes make things as they are. Genes seem to operate in the same way as other essences, whether these be the four humors, mana, or our souls. Moreover, as discussed in Chapter 2, the little understanding that most people do have about genetic mechanisms is that they function like switches—

that individual genes determine traits on a one-to-one basis, such as believing in entities such as "the height gene," or "the gene for depression." This simple—yet almost always incorrect—story is so much more appealing than the complexity of web-thinking. Furthermore, switch-thinking is a perfect match for essence-based views. It is easier to imagine a one-to-one correspondence between an essence and a trait than to consider multiple networks of elements that interact with your experiences. Switch-thinking goes hand and hand with essentialism.

## Genetic Essentialist Biases

Despite the appeal of essence-based explanations, we are still perfectly able to think that events happen because of external causes too—we often blame poor parenting for a badly behaving child, and we can see that a change in the tax law may cause people to save more money for retirement. We regularly consider such kinds of environmental or situational causes whenever they are salient to us. But you might say we have *essentialist biases*—once the idea of an essence is brought to mind, it is very hard to resist the notion that anything other than that essence is having much of a causal influence. It is as if we are holding our ear up to the world to listen for causal explanations, and the essence-based explanations resonate so loudly that they drown out other explanations.

Because genes are such apt essence placeholders, whenever they come up, our essentialist biases are activated, and they launch a train of particular cognitions. My grad student Ilan Dar-Nimrod and I identified the existence of four key genetic essentialist biases that influence our thinking when we encounter genetic ideas.[29] Each of these biases reflects ways that we think of genes in the same ways that we think about essences. First, when we hear that genes are relevant to a human trait, we start to think of that trait as *immutable*. We are more likely to think that nothing can be done to alter the trait, and that the outcomes are inevitable. Say, for example, that you learn that your partner has a loop of 48 nucleotides that repeats seven times at exon III on his or her DRD4 gene. This particular polymorphism has been labeled by the media as the "infidelity gene,"[30] and those who possess it have been shown to be more likely to cheat on their partners than those people whose loop of 48 nucleotides repeats fewer than seven times.[31] So upon learning that your partner has this particularly damning string of nucleotides, you may worry that your partner can't stay faithful to you if infidelity is wired right into his or her genes. Your partner's behavior may no longer seem fully under his or her control, and infidelity starts to sound preordained.

A second bias is that we think of a genetic cause as the *ultimate cause*. When we engage in switch-thinking, we mistakenly assume that carrying the gene and having the traits associated with that gene (i.e., the phenotype) are one and the same. This bias leads us to downplay any other potential causes that can shape the phenotype— the pertinent genes are perceived to be *all* that matters. The genetic test itself then can feel like a diagnosis—if you have the relevant gene, you believe you are destined to get the associated condition, whereas if you don't have the relevant gene, then you are assumed to be free of the condition. So if your sister got herself genotyped and discovered that she doesn't have "the gene for breast cancer," she may conclude that she no longer needs to get screened with regular mammograms, whereas what the negative result really means is that your sister's chances of getting breast cancer are only slightly lower than the population average, and that mammograms would still be about as useful for her as for anyone else. We'll discuss genes and disease risks more fully in Chapter 4.

A third bias is that we think groups that share a genetic basis are *homogenous* and *distinct* from other groups. Because we think of essences as carving nature at the joints, when we hear about how genes are shared by groups, then members of those groups come to seem more similar to each other and at the same time more differ-ent from other groups. So if we reflect upon the notion that African Americans have a shared genetic heritage, then African Americans may start to seem more similar to each other; we are more likely to assume that a trait possessed by one African American would be shared by other African Americans. And conversely, we might assume that a trait possessed by an African American would be less likely to be shared with someone of a different heritage. The genes can seem like fences that divide up the world between "us" and "them."

Moreover, this tendency to dwell on the genetic boundaries between groups interacts in pernicious ways with our tendency to view genes as ultimate causes. It can result in the mistaken belief that genes *always* underlie group differences. Groups can differ from each other in many ways, and genes may well be involved in some of those differences. But it is a logical fallacy to assume that because genes are involved in *some* differences between groups, *any* group difference must then also rest on a genetic foundation. Yet our essen-tialist biases, working in concert, are powerful enough to conceal this fallacy. So, for example, if people notice that African Ameri-cans are less likely to graduate from college compared with White Americans, they may well assume that underlying genes must be the ultimate cause of this racial difference—that is, our essentialist biases play a key role in sustaining racist ideas, which we'll return to in Chapter 7. Thinking of genes as essences makes it seem that people from particular groups are all painted by an identical genetic brush, and this same brush then gives rise to all their distinctive characteristics.

Last, we think of genes as *natural*—they seemingly came directly from God, and when we act in accordance with our genes, we're acting in the ways that we think we "should." We often automatically equate natural properties with ethical ones; things that seem natural are thought of as inherently good, and things that are unnatural can seem intrinsically bad.[32] We can see this fallacy in operation in the way that many products are labeled. The label "all natural" or "only natural ingredients" is typically used as shorthand to mean "healthy and good for you and the environment." There's no real need to read the label carefully to see what precisely is included—it's all good stuff. Often natural products really are good for you, but of course, not all that is natural is healthy for humans—think of poisonous mushrooms or hemlock, or of encountering a grizzly bear on a hike in a pristine wilderness. Yet because we think of genes as arising from natural processes, the mere mention of genes can prompt this naturalistic fallacy. For example, when people hear a discussion about "genes for homosexuality," this highlights the notion that homosexuality is natural, and it can immediately become much more acceptable to someone with socially conservative views. Likewise, we can see the naturalistic fallacy operating in reverse when we think of genes that have been artificially manipulated in genetically modified organisms (GMOs). The very idea of altering an organism's genome strikes many as being about the most *un*natural thing that they can think of. And in accordance with the naturalistic fallacy, people will often conclude that this makes such kinds of GMO products both morally wrong and potentially dangerous.[33]

Whenever people consider how genes might be relevant for something in their lives, they begin to think of essences, and these four genetic essentialist biases are launched. In cases of strong genetic causation, these genetic essentialist biases really aren't biases at all but are accurate reflections of the underlying condition. For example, in the case of FOP, the condition that Harry Eastlack was afflicted with, we can see that it is largely immutable, only people who have a specific mutation in the ACVR1 gene develop the disease, those who share this mutation are similar to each other in their prognosis and are different from those who do not, and the condition is indeed natural. It is rational to think of FOP in these essentialist ways.

Yet the simple switch-thinking essence-based story behind FOP is the exception and should not be generalized from. The vast majority of genetic influences on our lives operate in immensely complex ways and are not at all accurately captured by our essentialist biases. But even phenomena that are best understood with web-thinking have relevant genetic associations, and this means that our essentialist biases can be activated when thinking about them as well. Take the case of divorce. Research finds that divorce is substantially heritable, which means that genes are relevant for divorce in some unidentified ways. When we talk about the heritability of divorce, then we may start to think as though there were an essence underlying divorce. We may then be more likely to think of divorce as an uncontrollable prospect, that its emergence is almost exclusively caused by genes, that divorced people are somehow all alike and are fundamentally different from all nondivorced people, and that divorce is natural, and thus not as undesirable as it would seem otherwise. Divorce can really seem like a different phenomenon when we look at it through essentialist lenses, but it's important to remember that our biases are providing us with a distorted and inaccurate picture. Moreover, because the media enthusiastically report on the fascinating genetic associations that researchers around the world are routinely finding, coupled with the fact that the first law of behavioral genetics reminds us that almost everything is heritable and has genetic links, simply reading the daily news can trigger our essentialist thinking.

Assuming that the world is built upon a foundation of powerful and hidden essences is an irrational but highly intuitive way of thinking. Everyone thinks this way sometimes. Despite my own research, I still regularly think in ways consistent with the four genetic essentialist biases described earlier: for example, I think I would have the same emotional reaction to learning that my partner had the "infidelity gene" as would anyone else (though I'd hope further reflection would help me overcome the urge to act on the basis of my essentialist biases).

Being surrounded by imagined essences can seem rather threatening because they seem to deprive us of a sense of control and our freedom of choice. But what makes essences even more bothersome is that they so frequently get connected to some of the most controversial social topics that there are, such as mental illness, sexual orientation, and race. How do these topics change when we look at them through the lens of genetic essentialism?

# 4.

# The Oracle at 23andMe: Genetic Testing and Disease

THE MOST FUNDAMENTAL TRUTH OF LIFE IS that it always ends in death, or as Louis CK put it, we're all just dead people who haven't gotten around to dying yet. And typically, we die from just one particular cause, which remains unknown to us until it happens, so we spend our lives waiting to find out precisely how our own story will end. The anthropologist Ernest Becker proposed that this situation represents a debilitating quandary that is unique to humans; we are the only animal that is certain of our own impending demise. We know for sure that it's coming, but we usually remain poignantly unaware of the details of our ending, giving rise to much existential angst.[1]

The Jewish liturgical poem "Unetanneh Tokef" encapsulates this theme well. The poet notes that at the beginning of every year, it has been determined who will die, and of what causes.

*Who is to perish by fire, who by water,*
*Who by the sword, and who by wild beasts,*
*Who by hunger, or who by thirst,*
*Who by an earthquake, or who by the plague,*
*Who by strangling or who by lapidation.*

But what if we *could* find out the ending to our own life stories? Imagine you could visit an oracle who, peering into a crystal ball, gives you a list of all the possible ways that you could die, and gives you a percentage chance for each one. We all have a risk greater than zero that we will contract a whole host of deadly diseases—lung cancer, Crohn's disease, breast cancer, coronary heart disease—the list depressingly seems to extend on forever. And with the advent of direct-to-consumer genomics companies, you actually can now learn of your relative risk to these and many more diseases. The whole genetic testing experience can feel very much like meeting that oracle.

Consider the following case: In 2008, at the age of 48, despite not having any obvious health problems, Jeffrey Gulcher decided to swab his cheek and send in his DNA to be genotyped. When his results came back, he learned that he possessed fairly typical risk factors for most of the diseases that were screened for, but two of the results from his testing caught his eye: he had a higher than average lifetime risk both for developing Type 2 diabetes and prostate cancer. Being proactive about his health, Gulcher addressed his heightened diabetes risk by vowing to watch his diet and to exercise more. In response to his prostate cancer risk, which was 1.88 times higher than the average male, he opted to see his physician to get further tests done. His physician gave him a prostate-specific antigen (PSA) blood test, which yielded a PSA score of 2.0—slightly high but still in the normal range for a man under the age of 50. Typically, a 48-year-old man with a PSA score of 2.0 would indicate that the risk level for prostate cancer was rather low, and no further tests would be called for. But given the results of Gulcher's genetic testing, his doctor wasn't as sanguine about his relatively low PSA score, and he referred Gulcher to a specialist. Gulcher met with the renowned Northwestern University urologist William Catalona, the pioneer of the PSA test, and he gave Gulcher a more sensitive test for PSA, which revealed a score of 2.5, around the boundary of concern for someone of Gulcher's age. Given the PSA results and the findings from Gulcher's genetic testing, Catalona recommended that Gulcher receive a biopsy—an invasive and unpleasant procedure, in which an imposing eight-inch ultrasound wand is pushed deep into the rectum, and then a spring-loaded gun fires a dozen little needles into the prostate gland, removing tiny samples of prostate tissue. Gulcher agreed to have the biopsy test, even though Catalona's recommendation was an unusually cautious one. This proved to be a prescient decision. Despite appearing in good health, and having an initial PSA result in the normal range, the biopsy indicated that Gulcher had high-grade (Gleason score 6) prostate cancer on both sides of his prostate. Gulcher subsequently received surgery to remove the tumor (which further tests revealed was a more aggressive grade 7 carcinoma), and he was relieved to discover that the cancer had been caught before it had metastasized. The simple genetic test that Gulcher received, which is part of the testing performed by many direct-to-consumer genomics companies, may well have saved his life.

Jeffrey Gulcher isn't just an ordinary patient; he is the chief scientific officer and cofounder of one of the larger direct-to-consumer genomics companies, deCODE Genetics. The story of how his company's genetic testing may have saved his life has been told many times, including to the US Senate in 2009, and it is a powerful marketing statement about the potential health benefits of his company's services.[2]

A more famous case of the value of genetic testing comes from Angelina Jolie, who, in 2013, opted to have a preventative double mastectomy upon receiving her testing results that indicated that she carried a rare but highly dangerous mutation in the BRCA1 gene.

Given her family history and her genotyping results, her doctor estimated that she had an 87 percent chance to develop breast cancer. Jolie chose to give up her breasts, which showed no signs of cancer at the time, in return for the peace of mind that she was far less likely to develop the breast cancer for which she had heightened genetic risk. Jolie's decision also may have saved her life.[3] These examples are testimony to the striking health benefits that the genomics revolution promises.

Cases like Gulcher's and Jolie's can make us feel deeply unsettled about the genomics industry, though. On the one hand, it's reassuring that we now have affordable technology to identify genetic risks that standard screening measures miss. Who wouldn't be excited about the idea that a one-time affordable test could potentially identify the otherwise invisible loaded guns that are pointed directly at us *before* they've had the chance to be fired? But at the same time, these stories are terrifying because they completely upend the way that we think about health. Once upon a time, the question of whether or not you were healthy could be answered by a quick scan of your current state. If nothing hurt or itched, if there was no discomfort to note, and if you could find no blemishes or wounds, you could feel peace of mind in knowing that all was well. Alas, the advent of direct-to-consumer genomics companies like 23andMe has spoiled this state of tranquility. Because the one universal experience of being genotyped by 23andMe is that we *all* have some increased genetic risks to a number of diseases. Your own genotyping results will almost certainly reveal that you have a higher than average risk for at least some of the horrible diseases that are out there, and you likely are also a carrier of a number of recessive genetic disorders that may have fatal consequences for your children if you reproduce with another carrier. Rather than feeling at peace with the belief that we're in good health when no symptoms are present, genetic testing highlights that we're all potentially ill. It suggests that our essences are, at root, defective—diseases are engraved in our genes as probabilities.

## Thinking about the Genetics of Disease

If whether or not a person develops a disease is tied up with the particular arrangement of nucleotides along his or her DNA, how does
this knowledge affect the ways that we think about disease? Recall
from Chapter 1 the study that we conducted in which we had university students read an article that discussed either how genes influence obesity or how people's social networks affect obesity. Those
who read about the obesity genes subsequently ate more cookies than
those who read about obesity networks. The results of that and other
studies revealed that when people think of obesity as genetic, they
become more fatalistic about their weight.[4] If genes are involved in
obesity, then people are more likely to believe that there's nothing
they can do about it, and they're then not really responsible for being
overweight—it's their genes that should be blamed. A headline from
*The Onion* captured this fatalistic rationale well: "Obesity caused
entirely by genes, obese researchers find."

The effect of genetics on our thinking about disease is most evident for mental illnesses. Most mental illnesses, like pretty much
everything else, are heritable; they have a clear genetic component
underlying them. Because of this heritable basis, some people are
more genetically predisposed to develop a given mental illness than
others, whether that be autism, attention deficit disorder, alcoholism,
or virtually any kind of debilitating mental condition. Moreover, a
number of genetic variants have been identified that increase the likelihood that one will develop a mental illness. For example, a variant
on the 5-HTTLPR region of the SLC6A4 gene has been associated
with an increased likelihood of depression in response to traumatic
life events.[5] And over recent years, we have become increasingly aware
of the biological and genetic bases of mental illnesses.[6] How does
learning about the genetic foundation of a mental illness change our
thinking about those illnesses?

Believing that genes underlie mental disorders can change the very way that we make sense of what the disorder is. If, say, depression is viewed to have a genetic basis, then depression is seen as more of a biological problem than a psychological one. It reflects a defect in a person's physical body rather than in his or her thinking. And if this problem is grounded in biology, then we feel that it needs to be treated with a biological cure. We start to believe mental disorders should be treated with medications, rather than with psychotherapy.[7] This means that how we frame the causes of mental disorders doesn't just affect how we think about them—it can also shape the whole medical infrastructure that arises to most effectively treat them. Over the past few decades, there has been a striking rise both in the likelihood that medications are used to treat mental disorders and in the adoption of a biological model to explain those disorders.

Another change occurs in the way that we ascribe blame for such conditions. Often, when we have friends who are struggling with pronounced mood swings, substance abuse, debilitating anxieties, odd behaviors, or problematic thoughts, we might get frustrated with them—why can't they just get their act together? We might feel that our friends should just try to be more responsible, force themselves to confront their demons, quit their addictions cold-turkey, or seek professional help if it's beyond their individual abilities. The struggles that people have with mental illnesses can sometimes seem like moral failings—the afflicted are not living up to the standards that we expect of them. However, when we learn that someone's condition has a genetic basis, the condition no longer seems to arise from a person's poor judgments, lack of self-control, or bad habits—it now is grounded in a deep and inviolate essence. And when most people reflect upon this imagined essence, they are less likely to dole out harsh judgments or blame to someone who is battling a mental illness.[8] It no longer reflects upon the moral character of the afflicted. Moreover, a genetic basis of a psychopathology suggests that the condition itself is natural, and as the naturalistic fallacy dictates, it thus can't be all that bad. It's just part of who he or she is.

If people are blamed less for their mental illnesses when genes are implicated, then this has an additional benefit: genetic conditions don't seem as shameful or as stigmatized as conditions without an accessible genetic attribution.[9] This is extremely encouraging because one of the most damaging effects of suffering from a mental illness is the perceived stigma of the illness. The shame of the illness often leads people to isolate themselves, which prevents them from receiving the benefits of social support, and their well-being can spiral downwards. The perceived stigma can also lead people to deny to themselves that they have a problem or that they need to seek any professional treatment.[10] Hence, reflecting upon a genetic basis of a mental illness can have the positive consequence of making people feel less ashamed and more proactive about seeking treatment for their disorder.

However, our essentialist biases frequently cut both ways, and there are also some potent downsides to thinking about mental illnesses as having a genetic basis. First, because we think that essences carve nature at its joints, learning that there are genes associated with a condition heightens the perceived distance between those who are afflicted and those who are not. If you know someone with a mental illness, then reflecting upon the genetic foundation of the condition is likely to make you think of him or her as really different from you—you don't share the same fundamental essence. It might even seem as though the person were of a different species.[11] And if people think of the mentally ill as fundamentally different, then they'll also likely view them as more unpredictable, harder to relate to, and potentially frightening. This is particularly the case with schizophrenia, a condition often perceived as threatening. When people focus on the genetics of schizophrenia, they start to view people with schizophrenia as more dangerous, and they make more efforts to steer clear of them. Even though someone's mental illness may no longer seem as though it's his or her fault if genes are implicated, this doesn't mean that the individual will be viewed as harmless. If people think of someone as having a damaged essence, they very well might do their best to avoid the person.

Reflecting upon the genetics of mental illnesses also impacts our view of the prognosis of the illness. Imagine that you've been struggling with some symptoms of depression, such as feelings of sadness, futility, or experiencing a loss of energy. This is not at all an unlikely occurrence: as many as 17 percent of Americans get a diagnosis of major depressive disorder at some point in their lives,[12] and almost everyone occasionally experiences some symptoms of depression in the face of life's challenges. Now, consider how you might respond to learning about the existence of "the depression gene," which is what the 5-HTTLPR region is sometimes called. Your moments of darkness might not feel so fleeting anymore—if you're feeling down, maybe the depression gene is the source of your problems. And if you do have the depression gene, then what hope do you really have of ever feeling better again? After all, you're always going to have the same genes—they're part of your essence—so you may feel permanently cursed.

One of the most pernicious challenges in coping with depression is that when you are in a depressed mood, it's extremely difficult to remember what it was like when you weren't feeling depressed. The memories that most easily come to mind are all your failures, rejections, hardships, losses, and humiliations—the depression comes to feel like a rational response to a life that is so obviously full of misery.[13] Genetic thinking exacerbates this effect: not only is it difficult to remember the good times, but now you have the expectation that your disease, like your genes, will always be with you. Research finds that when people reflect upon the genetic essence of mental illnesses, they not only become more pessimistic about their own prospects but also about the prospects of their ill friends.[14] These pessimistic views of mental illnesses are a striking cost of genetic essentialism, as one of the key steps in the recovery process is being able to rekindle a sense of hope. Without the feeling that the mentally ill can tackle their conditions, they're less likely to take the proactive steps necessary to battle their illness.[15] A belief in the genetic foundation of a mental illness can thus severely undermine any attempts for recovery.

It's important to note, however, that all these kinds of genetic essentialist responses are really not justified because of the ways that genes influence mental illnesses. The so-called "depression gene," 5-HTTLPR, is only weakly linked with depression, and it's been connected to so many other phenomena, both positive and negative, that it's really hard to say with any precision how this gene affects our mental states.[16] The only evidence that exists for the link between this gene and depression is that some studies have found that people with the harmful variant of the gene are more vulnerable if they have also suffered through traumatic life events. But even this qualified finding is notoriously weak and hotly disputed.[17] It appears that there are no simple genetic switches for mental illnesses—like almost everything else, mental illnesses arise out of a complicated web of biological and environmental influences that unfold across one's development. Thinking of mental illnesses as essences is simply the wrong way to go about understanding them.

## The Trials and Tribulations of Genetic Testing

These biases persist, though, and are fueled by the almost daily enthusiastic news coverage about the most recent genetic discoveries. Moreover, as Jeffrey Gulcher and Angelina Jolie have shown us, we now have the opportunity to learn our own genetic risks for diseases. Recently, there has been a proliferation of more than two dozen direct-to-consumer genomics companies, many of which will provide results about your genetic risk for hundreds of diseases, sometimes without any consultation whatsoever.[18] And with the enactment of the Genetic Information Nondiscrimination Act in 2008—which makes it illegal in the United States for employers and health insurance companies to discriminate based on genetic information—there are fewer potential harmful consequences associated with getting your genes tested (although it can still be more costly to get life and disability insurance if you are at higher genetic risk for a serious disorder). With more genetic testing options, and less cost to the consumer than ever before, would you want to know what kinds of fatal flaws may be interwoven into your own genome?

This is a decision that could very well change the way that you think about yourself. Our genetic essentialist biases make us prone to view a disease-risk gene as an actual diagnosis for that disease. Even though Angelina Jolie had yet to show any signs of the disease, her genetic test seemed to have all the clairvoyance of an oracle—it could look past all outward signs of a young and healthy woman and magically conjure up the diseased essence that invisibly lay below. In the face of something as all-seeing as a genetic test, it would seem that the natural response would be to bow low and tremble in fear as you wait to find out what this oracle might pronounce. But is this an accurate description? Is getting genetic testing really akin to meeting an oracle?

Much media coverage suggests that it is: if you recall the discussion from the first chapter about Elvis Presley's DNA from the British program *Dead Famous DNA*, you get the impression that Elvis's DNA contained virtually *all* the answers to his life. The show's host, Mark Evans, and his team of geneticists were able to sleuth around in Elvis's DNA to discover the key markers that revealed that Elvis was destined to have glaucoma, frequent migraines, an expanding waistline, and ultimately, the flawed heart that killed him. *Dead Famous DNA* may be more sensationalist than the typical way that genetics is presented in the media, but its underlying switch-thinking message is not unusual. Each new genetic marker discussed in news accounts is often described as though it were the main, or only, predictor of a disease. The genetic variant that Angelina Jolie possesses (BRCA1) is frequently called "the breast cancer gene"—a label that directly activates the essentialist bias that genes are ultimate causes and akin to diagnoses—even though variants on this particular gene are only involved in about 5 percent of breast cancer patients.[19] But if genes really operated the way they're so often portrayed, then genetic testing would really be like visiting an oracle—you would learn which diseases are already lurking within you. The experience would be terrifying.

This is how many of us view genetic testing. I've been struck by how many people tell me point-blank that they would never consider getting tested themselves because they're simply too anxious about what they might find out. Among my fearful friends are several psychologists who specifically study human psychological biases and risk perception—the kind of people who, you would think, would largely make rational decisions and would be less impacted by any irrational fears or biases in their decisions. Well, that's simply not true—some of the worst decision makers I know are people who study decision making for a living (I guess there's some truth in the maxim that people often choose to study the questions that they don't understand themselves). But people who are afraid of getting genetic testing are not just limited to psychologists—one online survey found that 40 percent of people imagined that they would feel worried and anxious if they got genetic testing.[20] When the journalist Camilla Long wrote about the prospect of widespread genetic testing, she posited that the tests would provoke widespread panic attacks, concluding that "the cost to our collective mental health is incalculable." She questioned "who—in their right mind—would want to know? . . . The impact on one's life of that kind of Damoclean diagnosis is almost impossible to imagine."[21]

With that lurking in the back of my mind, I decided to get my own genes tested to explore my own reactions.

I was tested in three different ways, with three different direct-to-consumer companies. First, I had the haplotypes (a group of genetic markers that are inherited from one parent) on my mito-chondrial DNA and my Y chromosome analyzed by a company

called Genebase. This analysis only tells you about the likely ancestry of your maternal and paternal lines of descent, and I'll discuss this process in Chapter 6. Second, I was genotyped through
23andMe, which provides users with data for about 650,000
single-nucleotide polymorphisms (SNPs). Although this is just a
small fraction of our genetic code (remember the whole code is
about 6 billion nucleotides long), these are the places along the
genome where people are most likely to differ from one another. This
test provided information both about health risks and about ancestry.
Third, I had my full exome sequenced by a Belgian company called
Gentle. The exome is the part of the genome that codes for proteins,
and any mutations found in the exome are considered to be a more
likely risk for disease than mutations in other parts of the genome.
Full exome sequencing is a much broader sampling of the genome
than that provided by the SNP analyses of 23andMe, although there
is much overlap in the specific regions of DNA that are covered by
the respective tests. Gentle's service, however, does not provide any
information regarding ancestry. Gentle's full exome sequencing
was by far the most expensive service of the three (it cost $2,000,
which was about five times more than what Genebase and 23andMe
charged me at the time), but it included a thorough interview with
a genetic counselor, whereas the other companies only provided me
with information online. None of these tests was at all invasive—I
was merely required to either spit into a tube or scrape my cheek with
a swab. As is probably already clear to you, I feel quite strongly that
genetic testing is something that, in almost all cases, people should
really not get themselves that worked up about. Although it can be
incredibly tempting to engage in switch-thinking, we need to always
keep in mind that our thousands of interacting genes are just one
part of a complex biological system that is continuously interacting
with our experiences. I am so committed to this perspective that I
decided to write a book about it. But I'm still human, and I'm absolutely still prone to essence-based thinking.

When I received the results from my own genotyping from 23andMe, I made sure to first mix myself a stiff cocktail, take some deep breaths, and steel myself as I opened up my results for genetic health risks. And I'm glad I had that cocktail. Because if genes actually were the ultimate causes of diseases, I'd be in big trouble. According to 23andMe, I have "the genes for" psoriasis, prostate cancer, chronic kidney disease, melanoma, Parkinson's disease, ulcerative colitis, multiple sclerosis, narcolepsy, stomach cancer, Hodgkin lymphoma, scoliosis, restless leg syndrome, male breast cancer, scleroderma, hypothyroidism, esophageal squamous cell carcinoma, hay fever, chronic lymphocytic leukemia, and high blood pressure. And these are just the elevated genetic disease risks for conditions that 23andMe rates as being grounded in solid scientific evidence—the list is much longer when 23andMe also includes conditions that scientific consensus rates as being more iffy. I had to keep reminding myself to avoid switch-thinking, but these heightened risks still made me anxious. My results also revealed that I'm a carrier for two rare recessive disorders—alpha-1 antitrypsin deficiency and phenylketonuria—meaning that if my partner was a carrier for these same disorders, then each of our children would have a one out of four chance of inheriting these devastating conditions.

When I first received my genotyping results, I had just learned from my mother that her sister had been diagnosed with Parkinson's disease. And there, staring right at me, was 23andMe's report that I had a 32 percent greater than average genetic risk for developing Parkinson's. Suddenly, it seemed that my aunt's fate and my own were bound by a common strand of defective DNA—a newly discovered family curse that seemed woven into my very being. Against my better judgment, I impulsively did an online search for the symptoms of Parkinson's to see if I had any early warning signs (so far, so good). And I had a milder version of this neurotic reaction for each condition that 23andMe listed as a higher risk for me—I found myself ruminating upon whatever kind of relevant evidence I could drum up. Was the sunburn that I got on my neck last year finally going to release my dormant genetically inscribed melanoma? Did my low blood pressure scores invalidate the genetic results showing me to be at risk for high blood pressure? Or had my doctor maybe measured it wrong? Did I need to go back to get it reassessed?

Of course, not only will genotyping reveal that we all possess some elevated genetic risk factors for many diseases, but it will also show that we possess *decreased* genetic risk factors as well. In my own case, 23andMe tells me that I'm at less than average risk for celiac disease, both Type 1 and Type 2 diabetes, venous thromboembolism, age-related macular degeneration, Crohn's disease, pancreatic cancer, Lou Gehrig's disease, nicotine dependence, migraines, pulmonary fibrosis, kidney cancer, and a whole host of other diseases that I had never heard of before. You might expect that this is the kind of good news that would lead me to pump my fist in celebration. But, curiously, I didn't have much of a sense of elation or relief upon viewing this long list of good news, and my reaction is evidence for another kind of psychological bias that is common to the experience of getting genetic testing: negativity dominance.[22] Simply put, this bias indicates that at a psychological level, bad is stronger than good.

A good way to reflect upon the power of this bias is through a thought experiment: imagine that you visit the vineyards at Châteauneuf-du-Pape in France and are given a sample of the exquisite Clos St. Jean, right from the barrel. It's as heavenly as you might expect. Now imagine that you watch as a teaspoon of raw sewage is poured into the barrel of wine and is stirred in carefully. Would you be willing to have a second glass? My bet is that this tiny dollop of sewage is enough to make you gag at the thought of drinking the wine—it has thoroughly ruined the whole barrel. Now imagine the reverse situation: You see a barrel that is full of raw sewage, and you're overwhelmed with just how intensely disgusting the smell is. And someone then adds to the barrel a teaspoon of Clos St. Jean, and stirs it in. In fact, have the person pour in the whole bottle; it won't matter how much is added—the sewage is just as disgusting as before. The wine does nothing to improve it.

Similarly, a little bit of bad news will usually have far more psychological impact on you than will an equivalent amount of good news, and this bias has been documented across a wide variety of situations. Research finds that we remember bad events longer than good ones, we're more sensitive to losses than to gains, bad events affect the quality of our day more than good ones do, we judge people more by the bad acts they do than by their good acts, the success of marriages hinges more on the number of bad interactions than on the number of good ones, and, in general, the brain appears to be wired such that it responds more intensely to negative over positive information.[23] For example, if you learned that someone had once murdered someone else, you would obviously form a distinctly negative impression about that person. Well, suppose that murderer later risked his life to save someone else's life. Would that cancel things out? Would the new balance sheet indicating a net zero for lives lost and gained make you view him as positively as you do most other people, who also usually share a similar balance sheet total? Probably not—the one murder weighs more heavily on your mind than the one saved life. So how many individual acts of life-risking heroism would a onetime murderer need to perform in order to be forgiven for the previous murder? A large class of American undergraduates that was asked this precise question revealed that the median estimate was 25 lives—25![24] The moral credit that you earn for saving a life is only about one twenty-fifth of the moral condemnation that you bear for taking a life. Bad is indeed stronger than good.

When getting genetic testing, the psychological potency of negative dominance is brought into sharp relief. Each disease that you discover your heightened genetic risk for suddenly gives you something new to worry and obsess about. But there's not a symmetric sense of relief in finding out about your decreased risks. It's not like I had been worrying before about possibly getting, say, Lou Gehrig's disease. I have never heard of anyone in my extended family who had ever had this, and the disease has been completely off my radar. Learning about my lower risk for Lou Gehrig's disease didn't ease any existing worries because I hadn't been worrying about it in the first place. So even if you learn, for example, that you have 30 instances of diseases with increased genetic risk, and 30 instances of diseases with decreased genetic risk, it's unlikely that psychologically you will feel that it all balances out in the end. The heightened risk factors will probably weigh on your mind more than the decreased risk ones, and whenever you encounter information that is consistent with those heightened risk diseases, like learning of an afflicted relative, or detecting something in your body that might be consistent with a symptom of the disease, your sense of worry will likely shift into a higher gear.

## Why We Are Frightened by Genetic Testing: Mispredicting the Future

I've described my experience—how are others reacting when they learn of their genetic risks? Recently, a growing number of studies have been conducted to examine the psychological consequences of getting genetic testing. Participants are contacted several months to a year after they receive their testing results, and they are interviewed about their reactions. Are people having the panic attacks that my colleagues expected for themselves? By and large, no. The typical responses of people in these studies are that they are having far *fewer* anxious reactions to the results from their genetic testing than they had anticipated.[25] There were few instances of people showing any prolonged signs of increased anxiety because of their testing results— for the most part, their genetic risk information seemed to just roll off them like water off a duck's back.[26]

In stark contrast to the "incalculable" cost to our mental health predicted by some, and unlike my own roller-coaster experiences, genetic tests don't seem to burden people with significant worries or anxiety. Why is there such a divergence in people's reactions here? The optimistic conclusion from these studies needs to be offset with a couple of key caveats. First, in the largest study conducted to date, even though *on average* people who were tested didn't show many signs of anxiety, there was a subset of people (10 percent) who did report some distress, and the amount of their distress was correlated with how much heightened risk they faced[27]—that is, the ones who were most anxious were largely the ones whose genetic testing results gave them something to worry about. This suggests that although *most* people won't experience much stress as a result of testing (probably because most people won't get particularly bad news), it remains a real possibility that the results will be quite frightening to many. In addition, people who are more neurotic in general or who are more anxious about their health are also the most prone to have anxious responses to genetic tests.[28]

Second, because these studies were conducted months after people had received their genetic results, they are surely underestimating the degree of people's initial anxieties. Any anxiety that people may have felt upon seeing their genetic results should have been the most pronounced at the time that they first viewed their results. This was certainly the case for myself—my greatest anxiety was experienced just as I was first contemplating my increased risk conditions. When people are reporting on their anxieties a few months later, it's quite likely that their worries have largely passed.

I say this with some certainty because there's a great deal of psychological research that shows we are notoriously bad at predicting our emotional reactions to future events—what psychologists call affective forecasting. To understand this, imagine that something terrible happens to you: for example, you become a paraplegic as the result of a car accident. How do you think becoming paralyzed would make you feel, both initially and in the long run? It turns out we can imagine, quite accurately, just how miserable we would feel upon discovering that we had become a paraplegic. What we can't do nearly as well is imagine the *duration* of our feelings—typically, we overestimate just how long we would feel miserable after getting this kind of bad news.

One reason for this misjudgment is that we overly focus on the event that triggers our emotional reaction.[29] Getting paralyzed would certainly make a person feel miserable, but we forget that this event doesn't happen in a vacuum—many other life events will happen, and continue to happen, that affect our feelings. Fortunately, people who get paralyzed don't actually spend the rest of their lives just focusing on what it's like to be paralyzed—they also worry about whether they have enough money to get the roof fixed, or wonder what people will think of a new recipe at their dinner party, or ruminate about their boss's critical feedback—and these thoughts tend to dominate their emotional lives more than the fact that they became a paraplegic several years ago. We fear genetic testing because of the same incorrect assumption that any potential bad news we encounter, such as an increased risk of getting Parkinson's disease, will be the focus of our attention for the rest of our lives. We don't realize that the everyday drama of our lives will continue unabated.[30]

Another reason we inflate the lasting impact of our emotional reactions to life events is that we fail to appreciate the power of adaptation. When good things happen to you—you win the lottery, you get a promotion, you find a new partner—with time, you adapt to the new reality, the initial happiness you felt at the time fades, and you return, more or less, to a fairly similar level of happiness you had before your good news. It's the same for bad events—if you are paralyzed in an accident, get divorced, or learn that you have a debilitating disease—over time you adapt to this new reality and your happiness slowly returns to more or less where it was before. There might be some lasting changes—lottery winners indeed are happier than paralyzed accident victims, even years after these events occurred—but the point is that the gap in the happiness between these two groups gets much smaller as time passes because people come to adjust to their new situations.[31] Psychologists refer to this adaptation as the hedonic treadmill, where our expectations keep pace with the unfolding of our life experiences. But the key to this hedonic treadmill is that we can adjust our expectations to a new situation only once it is clearly understood. What we can't easily adapt to, on the other hand, is the unknown. Psychologically, uncertainty keeps us in a state of agitation.[32]

## The Curse of Uncertainty and Its Role in Genetic Testing

The notion that we can adapt to certainty much better than uncertainty is particularly relevant for genetic testing. Getting yourself tested can either reduce or increase your sense of subjective uncertainty. First, let's explore how genetic testing can reduce feelings of uncertainty. Consider the situation of Anneke Van Kirk. Recently married, Anneke began to notice some increasingly strange behaviors in her husband. He was often fidgety, he paced around the house and made jerky movements, and he would frequently show odd grimaces on his face. He constantly complained of dizzy spells, and he became quite accident-prone, once even setting himself on fire. Although he was a loving husband, sometimes he could be remarkably callous, erupting into vicious tirades and verbally belittling those around him. He once defaced Anneke's prized book collection for no apparent reason. He would become sullen and insular for long periods. He stopped bathing or grooming himself, and he became so out of it that he didn't even visit Anneke in the hospital after she gave birth to his daughter. As his condition worsened, he began to shake violently and had uncontrollable muscle spasms. Ultimately, he could no longer control his muscles, and he was confined to a hospital bed, where he died a few years later at the age of 55.[33]

Anneke's husband, the folk music legend Woody Guthrie, suffered from the ravaging symptoms of Huntington's disease. Huntington's is a horrific disease, and it has a very straightforward and well-understood genetic cause. Along the HTT gene, on chromosome 4, there is a series of three nucleotides, CAG, which codes for the amino acid glutamine. This series repeats itself over and over within the gene (e.g., CAG, CAG, CAG). For the vast majority of people, this series repeats itself fewer than 35 times, and for these people the number of repeats appears to have no biological consequence. For other people, however, the series may repeat itself anywhere from 36 to more than 100 times. When the repeating sequence is this long, the gene creates a mutant form of the huntingtin protein, which will almost certainly lead to Huntington's disease if the carrier lives long enough. The larger the number of repeating series, the more deleterious is the mutated protein that accumulates over the individual's life, and it ultimately begins to kill neurons throughout the brain. The number of repeating series in your DNA can be read like a clock (with 95 percent precision)—if you have 41 repeats, then you will develop Huntington's disease by 75 years of age; 45 repeats means that you'd face the disease by the age of 55; and with 50 repeats you would only be disease-free until your 50th birthday.[34] The disease emerges from behind the curtain precisely when it receives its genetically determined cue, and it does so regardless of what you do with your life— whether you vigorously exercise, avoid gluten, get regular check-ups, or play Sudoku puzzles daily is irrelevant. In stark contrast to the complex web of influences for most diseases, Huntington's is controlled entirely like a genetic switch. This is the rare case where you might as well be dealing with an oracle.

But what makes Huntington's even more psychologically torturous is that it's one of the few known diseases that is "truly dominant" in Mendelian terms—whether you have only one copy of the Huntington's gene or two, the outcome is the same. This means that if one of your parents has the Huntington's gene, you have a 50 percent chance of inheriting the mutated gene yourself, and because the disease is fully penetrant with no known cure, possessing this mutant allele amounts to nothing short of a death sentence. As a young man, Woody Guthrie watched his mother suffer through the same symptoms he had 30 years later, and two of his daughters, Gwendolyn and Sue, inherited the disease from him. The genetic status of three of his other children remains unclear, as they all died young in accidents, whereas his remaining three children, including his son, Arlo, the folk musician, appear to have inherited the normal allele. The Huntington's gene is one of those rare cases in which genes really do operate just like essences, with the gene being the ultimate cause of an immutable condition. Having a parent with Huntington's ensures that you suffer from the most debilitating kind of uncertainty imaginable—you have a 50/50 chance of either being entirely free of the disease or of certainly developing the disease yourself. To make matters worse, people with Huntington's disease typically don't show any symptoms until they've already reached the age where they have had children themselves. A positive diagnosis would mean that you have to face not only the prospects of your own disease, but your children's future risk for the disease as well. Huntington's truly lives up to its moniker as "the cruelest disease."

Under such brutal circumstances, would you want to get yourself tested if, like Woody Guthrie, you had a chance of inheriting your mother's deadly gene?

Nancy Wexler thought she would. When she was 23, she learned that her mother had Huntington's, and thus had to confront the grim reality that both she and her sister were also pinned beneath their own Damoclean sword. She dreamed of being able to get tested for the disease to remove the haunting uncertainty that hung over her head. Despite a complete lack of formal training in genetics (a required introductory biology course as an undergrad was the extent of it), Nancy devoted her life to understanding the genetics of Huntington's. She focused her efforts on the shores of Lake Maracaibo

in Venezuela. The region had experienced a perfect storm of isolation, a history of inbreeding, and the fact that one of the region's
initial founders was a highly fecund woman, with the too appropriate name of Maria Concepción, who carried the lethal mutation.
When Nancy Wexler visited the region in 1979, she found the largest
known population of Huntington's carriers in the world: 371 people
with the disease, and another 3,600 at-risk individuals. And it was
in the blood samples of Maria Concepción's descendants that Nancy
Wexler discovered their family curse.[35] She identified the repeating
string of CAG nucleotides that caused the disease, and this led to the
first predictive test for Huntington's. Before the test existed, surveys
revealed that approximately two-thirds of people with biological relatives afflicted by Huntington's expressed interest in being tested. But
once an accurate test actually became available in 1993, the dramatic
psychological stakes of the test became overwhelmingly apparent.
Since then, only about 10 percent to 20 percent of at-risk individuals have gotten tested.[36] Even Nancy Wexler herself decided against
it, saying "I know that with me, if I were to go to bed every night
thinking, I'm going to die of Huntington's, you know, why should I
bother even getting up?"[37]

Because there's no known cure for Huntington's, the test only provides one with knowledge—an exceedingly brutal and unforgiving
kind of knowledge. Without a doubt, the Huntington's test reduces
your uncertainty. Does this mean that people who are at risk for
Huntington's would be psychologically better off if they got themselves tested? Individuals who choose not to be tested tend to be
more pessimistic about their outlook than those who go ahead with
the test, perhaps because they're already detecting early signs of the
disease or because their relatives had more severe symptoms.[38] Choosing not to go ahead with the test allows everyone to hold on to some
hope that they have dodged the genetic bullet.

But remember that uncertainty is the one kind of situation that
people are unable to fully adapt to. Given this, we would expect that

people would be better off having an answer—*any* kind of answer—
than living their lives in a perpetual state of limbo. One classic study
tested this precise hypothesis with Huntington's disease.[39] In the
early years of testing for Huntington's, the test was not definitive—it
frequently produced inconclusive results that conveyed no information to the participant about their own risk for the disease. The study
took advantage of this flaw in the test to explore the psychological
consequences of getting tested. All of the participants in this study
volunteered to get the Huntington's test, but approximately one-
third of them received inconclusive results. The other two-thirds of
the participants did get conclusive results, and approximately half of
this group received bad news. They found out that there was a very
high likelihood that they possessed the mutant allele, and thus would
someday develop the disease. In contrast, the other half received good
news, and found out that there was a very low likelihood that they
had the mutant allele, and thus would never have to worry about it
anymore. How did the participants respond to these test results? A
short-term follow-up about a week after the test results revealed that,
unsurprisingly, those who had received good news were far happier
than the recipients of bad news. But another follow-up, conducted
one year later, found that the happiness of those who had received
good news had largely evaporated over time, and the depression of
those who had received bad news had also on the whole faded with
time, such that these two groups reported being comparable in terms
of their well-being one year later. However, people in the group that
never received conclusive feedback remained in a bad state a year
after they had taken the inconclusive test; they were significantly
worse off in terms of their well-being compared with those who had
received conclusive news, good or bad.

Having a sense of certainty, good or bad, becomes a fixed point in
your psychological reality. It grounds you so you can begin to understand your situation, and then slowly, over time, come to terms with
this new knowledge. In contrast, not knowing bears a constant psy-

chological cost—you can never find true peace if you are just waiting for it to be your turn. Of course, it's important to note that this study was exploring how people responded *on average*, and there's always the very real possibility that some people will not show this more typical pattern of adaptation to the news of their test. Moreover, there is a possibility that some people might get so traumatized by the bad news of a test result that they do something drastic, such as commit suicide, although that didn't happen to any of the participants in the above study. Overall, though, for those rare conditions in which your genes are highly predictive, testing can provide a sense of certainty, which allows you to adapt to the reality of your situation.

But what about conditions in which your genes are *not* highly predictive? Remember that the vast majority of diseases emerge from complex webs of influences, both genetic and otherwise. In these cases, genetic testing only provides you with very limited probability estimates. Unlike the test for Huntington's, genetic testing for most diseases does not tell you with any certainty whether or not you're going to contract the disease.

It's important to keep in mind just how inherently uncertain the results are from direct-to-consumer genomics companies, and this uncertainty weakens the utility of the tests. One reason that these results are subjectively uncertain is that the tests provide information in the language of probabilities. For example, my genotyping results indicated that I have a 32 percent increased chance of developing Parkinson's. The problem with this kind of reporting is that we have a hard time understanding probabilities—it's not at all an intuitive way of thinking. Students frequently struggle with it. Even medical doctors sometimes grossly misinterpret probability information.[40] In a 2012 study, some people couldn't make heads or tails out of their genetic testing results. For example, one participant who was informed that she had a 30 percent increased risk for a particular condition responded by saying "I believe that means I'm 30 percent more likely than other people, or am I three times more likely than other people? That particular number means very little to me."[41] Other studies have shown that people sometimes simply reduce these probability estimates to a binary format—that is, they end up thinking that it's equally likely that they will or will not develop the disease, which is not at all true for most kinds of genetic risk.[42] The inherent complexity of probabilistic thinking means that receiving risk information can very well increase a person's subjective sense of uncertainty. For example, prior to getting my genetic test, I hadn't given much thought at all to prostate cancer. But now that my genotyping results shows that I'm at increased genetic risk for it, it's like a mental spotlight is shining directly on it, and whenever I encounter the words "prostate cancer" I'm now reminded that it's something that I'm at increased risk for. Still, my increased risk is only a highly nebulous probability. Probabilities speak to the average of a sample and have little to say about any given case. My own case—the one case that is most meaningful to me—remains as uncertain to me as it was before my genotyping, but my test results have brought this particular uncertainty to the forefront of my mind.

The uncertainty surrounding genetic disease risk is often compounded because the probabilities that most people have for contracting many of the diseases covered in genetic testing fortunately tend to be extremely small. For example, my 32 percent increased risk for developing Parkinson's disease sounds significant, but this means that my absolute risk for developing the disease is 2.1 percent rather than the population average of 1.6 percent. Is that a worrisome level of risk? I very well could draw a losing lottery ticket and develop the disease, but so could anyone else, and my odds really aren't all that different from average. The perceived utility of genetic testing drops when dealing with probabilities that are far from certain. For example, one study found that when women seeking genetic testing for breast cancer learned that the tests were not highly predictive, they became much less interested in getting the test.[43] We want an oracle who can speak with confidence, not one that can only speak to us in weak probabilities.

The genetic risk information conveyed by direct-to-consumer genomics companies is often at a rather trivial level because the link between each gene and each associated disease is typically so vastly small. The web of influences on most diseases is almost unfathomably complex. Take my Parkinson's test result, for example. I'm at increased genetic risk for developing the disease, but the vast majority of the factors that predict Parkinson's aren't covered by the genetic test. On average, only about one-quarter of the variance of developing Parkinson's is genetic, and genetic tests only provide risk estimates for a fraction of that genetic contribution (although depending on the specific mutations involved, the predictive validity can get much higher).[44] So this often leaves the consumer in a position not unlike that of a tourist going on vacation to a new destination who is trying to decide whether to pack an umbrella, with the only information available being a weather forecast from an issue of last year's newspaper. That one forecast is perhaps better than having no information at all, but it's of very little predictive value.

Although the identified genetic risks for the vast majority of diseases are so small that genetic testing is of little use, there are exceptions. My grandfather suffered from one of these exceptions, and it's engraved into most of my memories of him. I've always regretted what I did in one of these memories—I wish I had just refused my grandfather's request to borrow my bicycle. But I was only 14 at the time, and when he said he wanted to show me that he could still ride a bicycle, I didn't feel I could refuse. Trouble is, he was 80 years old, and had been suffering for years from Alzheimer's. I reluctantly lent him my bicycle and watched as he rode off down the street, proving that it's true that you never really forget how to ride a bike. Until he tried to get off, that is. His foot got stuck on the top bar and he fell hard in front of me, breaking his arm. But the next day he walked up to me proudly, held up his arm and said, "Look! It's better already!" "But Grandpa," I said, pointing to his cast, "it's your *other* arm."

The look of despair in my grandfather's eyes when he glanced down at his broken arm, seeming to realize just then how lost his mind already was, has been forever etched into my memory. I've always worried that someday my own eyes would show that same sense of anguish, and my grandfather was very much on my mind when I first went to open my genotyping results from 23andMe. This is because the strongest single genetic predictor of a common disease is the relation between the APOE gene and Alzheimer's risk. As the medical geneticist Robert Green described this particular genetic risk, "It is the only one that meaningfully increases your risk in a way that could conceivably mean something to an individual."[45] Those of European descent who have two copies of the $\varepsilon 4$ allele of the APOE gene have about a nine times greater risk to develop Alzheimer's than average (the correlations between genetic variants and Alzheimer's vary by ethnicity, as is the case for most other diseases as well).[46] This is an unusually large effect for a single gene and a common disease. To make matters worse, at present there's no known way to reduce the risk of Alzheimer's or to change its course. So it's a very big deal to find out your APOE status, and to signal the magnitude of this event, on 23andMe you first need to read a disclaimer and unlock a protected file in order to see which APOE variants you have. Two of the first few people to have their entire genomes sequenced, the co-discoverer of DNA James Watson and the psychologist Steven Pinker, both asked specifically to have their APOE status deleted from their genetic output—the only deletion that either of them requested. "I figured that my current burden of existential dread is just about right," Pinker told the *New York Times* about his decision.[47] Was I going to regret for the rest of my life that I had unlocked my APOE status?

The extent of my grandfather's dementia in his last years had been so extreme that I always assumed it must have had a solid genetic basis—I naively assumed that the genetic risk must be proportional to the severity of the symptoms. So it was with great trepidation that I clicked on the link, knowing that I'd never be able to unsee the results again. But, much to my pleasant surprise, I learned that I didn't have *any* copies of the APOE ε4 alleles. This was the single most significant revelation from my genotyping, and I certainly did feel some relief upon learning that I didn't possess the pair of harmful alleles. But, then, as I learned later, neither did my grandfather. My father recently got genotyped too, and he doesn't have any copies of the problematic APOE ε4 alleles either, so at most my grandfather could have only had one of these alleles. If he did have one copy of the ε4 allele, his genetic risk for Alzheimer's would be about 1.65 times higher than average—that is, a risk of about 12 percent instead of the population average of a 7.2 percent risk. Indeed, despite how relatively strong the link is between the ε4 allele and Alzheimer's risk compared with other common diseases, the vast majority of people with the ε4 allele don't develop Alzheimer's, and about half of people who develop Alzheimer's don't have this problematic allele.[48] For common diseases, the links between any single gene and the disease are quite tenuous—even the fearsome case of APOE and Alzheimer's is a far cry from a deterministic relationship.

The reason why the links between single genes and common diseases tend to be so weak is that any strong monogenic associations with particular diseases would lead to the genes, and their associated diseases, becoming less common in future generations because of natural selection—those who had the deleterious gene would be less likely to have surviving offspring. Instead, common diseases usually arise from a complex web of genetic, epigenetic, and environmental factors that emerge across a developmental period. They preclude any simple explanations and resist any simple tests. For example, the genetics of schizophrenia have been studied more than those of any other mental illness, and we know that the debilitating condition is highly heritable. Though only about 1 percent of the population develops schizophrenia, someone with a schizophrenic cousin has about a 2 percent risk, someone with a schizophrenic sibling has a 9 percent risk, and someone with a schizophrenic identical twin has a 48 percent chance of developing the disease.[49] However, despite this clear evidence of heritability, genome-wide association studies have estimated more than one hundred common genetic markers are associated with schizophrenia.[50] The *New York Times* journalist Nicholas Wade referred to the results of these studies as "a historic defeat, a Pearl Harbor of schizophrenic research."[51] With such a complicated network of very weak genetic links, it is hard to imagine that we'll ever be able to fully understand the genetic foundation of schizophrenia.

But the result of this study on schizophrenia is par for the course for genome-wide association studies—these studies typically reveal complex webs of extremely small associations that have thus far been of limited practical value for aiding our ability to treat the diseases.[52] Hundreds of these kinds of genome-wide association studies have been conducted, and their results are used by consumer genomic companies to estimate much of one's risk. But the risk information that is provided by these companies has attracted increasing amounts of criticism.[53] The geneticist David Goldstein referred to these kinds of risk estimates as "recreational genomics," as "the information has little or in many cases no clinical relevance."[54] The geneticist Thierry Frebourg claims, contrary to Jeffrey Gulcher's experiences with his prostate cancer risk, that "there is no evidence that 'positive' (DNA) tests, based only on the screening for common genetic variations, will justify a specific medical follow-up and procure a medical benefit to individuals."[55] Craig Venter, the key player in the initial sequencing of the first human genome, went so far as to call this all "useless information."[56] In some cases, the data from these genome-wide association studies have not been found to predict disease any better than old-fashioned reviews of a patient's family history.[57]

There's yet another problem with these genetic risk estimates. Because many genetic variants influence a condition, predicting an individual's risk requires you to somehow combine the disease risks across those different genetic markers. But how do you do that?

How could you calculate someone's risk, say, of Crohn's disease if the person has 11 of the 70 identified risk variants? It's not at all clear how the individual risks of multiple genes combine to influence the overall likelihood of developing a disease. Stanford medical geneticist Thomas Quertermous is more unequivocal: "Do we know how combinations of genes affect risk? The answer is completely no."[58] But this hasn't stopped direct-to-consumer genomics companies from offering customers highly specific risk estimates for complex diseases. For example, 23andMe calculates its risk estimates by multiplying all of your relevant gene's odds ratios together to produce a highly specific risk estimate for you for each disease.[59] But aside from the issue of whether it's correct to assume that each individual gene's risk factor is independent from all the others (which this calculation assumes), the challenge here is how to decide *which* individual gene risks to include in the calculations. First, different studies find different collections of genes that are involved in any condition—it's actually quite disturbing how little overlap there is between different genome-wide association studies. So depending on the direct-to-consumer company you use, the company will make strikingly different decisions about which genes to include in its tests.[60] Second, one reason different studies identify different genetic risk factors is that they're relying on different populations—the association between a given gene and condition is often very different among, say, Japanese than it is among Finns. Is it reasonable to include a risk estimate from a study that was based on a Finnish sample if your own ancestry is British? Moreover, often subsequent studies fail to replicate the genetic associations found with earlier studies, and it's not clear what the best way is to evaluate those earlier published associations.[61] Answering these questions involve some tricky judgment calls for which there is no obvious consensus.

Deciding how to combine the genetic risk factors across multiple genes is not just some technical statistical problem. It has dramatic consequences in the kinds of risk estimates that are produced. The depressing reality of genetic testing is that different companies will often provide dramatically different profiles of your disease risks.[62] For example, the *New York Times* journalist Kira Peikoff had genetic testing performed by three different companies; whereas 23andMe informed her that her most elevated risks were for psoriasis and rheumatoid arthritis, her results from Genetic Testing Laboratories indicated that these same two conditions were precisely the ones that she had the *lowest* risks for.[63] Similarly, the *Sunday Times* journalist Nic Fleming was informed by 23andMe that his risk for exfoliation glaucoma was 3.6 times higher than average, whereas deCODEme stated that his risk was 91 percent *below* average.[64] These kinds of conflicting risk estimates are not at all the exception.[65] One study compared genotyping results from different companies for the same individuals and found that not only did the different companies calculate a different risk estimate for each and every condition, but in *one-third* of these conditions, the results from the respective companies were in direct contradiction with each other—one company was predicting *increased* risk for a disease while another predicted *decreased* risk for the same individual.[66] This isn't much more agreement than you'd expect from visiting two different fortune-tellers. These different risk estimations are stark evidence that the direct-to-consumer industry is grossly overpromising valid results to their customers.

My own experiences with getting tested at both 23andMe and Gentle also provided me with conflicting information. Of the 100+ diseases that 23andMe specifically tests for, I was informed that I was at greater risk than average for about one-third of those, and I was also at lower than average risk for another third of them. My risk was typical for the remaining third. In contrast, Gentle's testing revealed that the number of diseases for which I had greater than average genetic risk for was a grand total of zero. Zero! Likewise, their testing indicated that the number of diseases that I have less than average genetic risk for was also zero. This seemed like a mistake—how could this be possible? When I asked Gentle's genetic counselor specifically about my heightened genetic risk for prostate cancer, Parkinson's, melanoma, and all the other conditions that 23andMe had flagged, she told me that the company felt that it is not currently possible to provide accurate risk estimates for these common diseases. Gentle identified the same genetic variants that 23andMe did—there was no apparent disagreement in the actual genetic results when the SNPs associated with the gene chip used by 23andMe overlapped with the same SNPs identified by Gentle's full exome sequencing. They were the same data—they were just being interpreted differently. Gentle realized that there wasn't sufficient scientific justification for offering any kind of specific prediction for these common diseases. What about the fact that I didn't possess any problematic APOE ε4 alleles? Shouldn't that mean I have a less than average likelihood of developing Alzheimer's, given the relatively strong predictive power of this allele? Gentle's counselor told me that for Alzheimer's, my case was simply an absence of bad news. And the company viewed an absence of bad news here as typical risk.

Clearly, the calculations made by 23andMe and Gentle provided very different test results. According to 23andMe, I have a lot to worry about—I have greater than average risk for a depressingly long list of conditions. But I also have much to feel relieved about, as I have lower risk for a long list of diseases too. But the much more exhaustive full exome sequencing provided by Gentle tells me that my risk for *all* common diseases is typical. These two companies read the same genetic text and told two completely different stories. It feels a lot like I'm caught between a Jehovah's Witness and a Catholic giving me conflicting spiritual guidance based on their respective interpretations of the Bible. And we do treat our genomes like a bible. We hope that our genomes hold some ultimate truths about ourselves, and we trust that direct-to-consumer genomics companies will reveal these essences to us. But the bottom line is that for common diseases—the diseases that are going to kill the majority of the people in the world—genes are nothing like essences. They are part of an inherently complex and interactive web of forces. As a result, precise risk estimates can't be provided, at least not with our current understanding. The oracle's crystal ball is made of mud.

You certainly don't hear this from most direct-to-consumer genomics companies. Many companies continue to advertise unrealistically optimistic health benefits of getting tested. One study by the Governmental Accountability Office found several "egregious examples of deceptive marketing" by direct-to-consumer genomics companies, including claims that the testing could identify which sport a child would be more likely to excel in, descriptions of risk predictions as though they were diagnoses, and sales of supplements that supposedly could repair their damaged DNA.[67] Reflecting on the state of the science underlying these predictions, the geneticist James Evans argued that "claims that DTC [direct-to-consumer] genetic testing represents 'a guide to your medical care' are simply false, and the laboratories that claim (or imply) that they are offering tests that are medically useful should be called out and prevented from doing so."[68]

Given this glaring disconnect between the imprecision and inaccuracy of the genetic risk estimates and the overly optimistic promises that several companies have made, it is perhaps not surprising that the Food and Drug Administration (FDA) stepped in, and in 2013 sent a warning letter to the biggest direct-to-consumer genomics company of all—23andMe—demanding that it stop providing health risk information in the United States. The FDA views the genetic feedback that 23andMe provides as a medical device. It is worried about the potential inaccuracy of both the genotyping (e.g., the test may say someone possesses the harmful variant of the APOE gene when it may have incorrectly identified that variant) and of the risk estimates (perhaps its calculations for predicting the risk of Alzheimer's are off). Moreover, the FDA expressed concerns about the possibility that people will misunderstand what their test results indicate and may make life-altering decisions, such as Angelina Jolie's preventative mastectomy upon receiving a potentially inaccurate test result for the BRCA1 gene. The FDA's ruling has been controversial; some maintain that access to information about your own genes is a universal right.[69] On the other hand, the FDA announcement was followed by a class-action lawsuit in which past customers are seeking millions of dollars in damages, arguing that the company's ads are misleading and that its risk estimates are "not supported by any scientific evidence."[70] In 2015, the FDA allowed 23andMe to provide information on whether or not people are carriers for 36 recessive disorders, such as cystic fibrosis and sickle cell anemia—these are all Mendelian diseases that are determined by a single gene. The company remains unable to provide health information for any polygenic conditions. At the time of writing, 23andMe has complied with the request, such that their new customers are not provided with any health information for common diseases with their genotyping. The future of 23andMe as a provider of health information is in question, although the company's vast and growing database of more than one million genotyped individuals, many of whom have also completed hours of questionnaires about various traits and symptoms, ensures that it will continue to play a valuable role in genetics research.

Given all this, should you get yourself tested? If you're looking for an oracle in your test results, you're most likely going to be disappointed. Aside from a number of rare conditions that have strong predictive genetic links, like Huntington's and rare kinds of breast cancer, you most likely won't get much clinical utility from the test. Although Jeffrey Gulcher's life-saving story makes for a moving marketing pitch, it's more of an example of a broken clock being correct twice a day than it is of the predictive precision of genetic testing for prostate cancer. My own genetic testing experiences provided me with some flashes of anxiety and relief when I contemplated all the highly specific risk estimates that 23andMe provided me. But I recognize that these reflexive emotional responses are entirely misguided; these risk estimates have very little predictive validity, which is underscored by the fact that different companies offer entirely different estimates. I'm certainly going to die of something, but my bet is that it's more likely going to be from something other than my higher-risk conditions identified by 23andMe.

On the other hand, people often choose to get genetic testing for the experience, rather than to seek medical guidance—in these early days of direct-to-consumer genomics, the most common reason that people offer for getting themselves tested is that it sounds like a fun thing to do.[71] And, indeed, it really is fascinating to look over your genetic profile—you get a new perspective of yourself, even if most of the predictions are not accurate—and it remains possible that future scientific findings might boost the clinical utility of your genetic testing results. I found the whole genotyping experience to be a lot of fun—akin to taking a BuzzFeed quiz. From a consumer satisfaction point of view, 23andMe really delivers, and in this respect the company outdid Gentle. 23andMe provided me with a gripping set of predictions about my health with real concrete numbers—I learned that I have a 2.1 percent chance of developing Parkinson's disease, and this is 32 percent higher than the average person. The 23andMe experience "felt" satisfying because it provided a wealth of highly specific and personal information about my health. But, then, so would the fortune-teller down the street, and at least she isn't claiming any scientific foundation to her predictions.

On the other hand, my results from Gentle, despite being far more scientifically justified, really didn't psychologically affect me much at all. A thorough sequencing of my entire exome, examined alongside the collection of 1,700 conditions that Gentle tests for, tells me that I'm nothing but typical. Before I took the Gentle test, I didn't know what diseases I should be watching out for, and after the test I still feel the same. In genetic testing, no news is good news, to be sure, but it still seemed like I paid a lot of money to Gentle for what in the end feels like no news.

Our essentialist biases lead us to expect that our fates are all inscribed in our genomes, and many direct-to-consumer companies promise to act like oracles by deciphering the coded text. But this is not at all accurate for the common diseases to which the majority of us will ultimately succumb. The excitement and anxiety that stems from genetic testing is the notion that you're dealing with an oracle who can invoke your underlying essence. But once you recognize that there are no precise risk estimates available for the most common causes of death, then the whole genetic testing experience becomes far less interesting. Direct-to-consumer companies are simply unable to accurately provide specific predictions about your health risks for common diseases because common diseases don't have a clear set of causes that can be read faithfully from your genes. The scientific reality of genetic testing is that there is no oracle behind the curtain.

# 5.

# Born This Way: Thinking about Gender and Sexual Orientation

IN THE '90S, ONE OF THE MOST POPULAR recurring characters on *Saturday Night Live* was Julia Sweeney's "Pat." The running joke was that Pat was so completely androgynous in every respect that it was impossible to tell what his or her gender was. The other characters in the skits would pose various indirect questions, hoping to be able to once and for all identify Pat's gender so they could charge him or her the appropriate price for a haircut or sell him or her the right kind of deodorant. So the other characters would haplessly scour for clues by asking Pat the name of his or her ex-fiancé/ée (answer: Terry), which magazine he or she would like to read (not *Sports Illustrated* or *Glamour*, but *People*), and what movie he or she would watch that evening (*Tootsie*). The skit was funny because the audience shared in the discomfort of the other characters. We understand just how unsettling it can be to not know someone's gender—in Germany, the government has even passed a law maintaining that a child's first name must identify its gender. This discomfort has deep roots—there is no social category of more significance than gender in helping make sense of those around us.

Upon your arrival into this world, the first statement that was probably uttered to describe you was to announce your sex; indeed, many prospective parents are so eager for this critical piece of information that they demand their ultrasound technicians tell them it months before their child is even born. The term "sex" generally refers to biological differences between men and women that encompass chromosomes, hormonal patterns, and sex organs, whereas the term "gender" refers to social roles—masculine or feminine—that a person may identify with. Although the notion that gender equates with social roles suggests that it is largely the result of one's experiences, most people think about gender as though it stems from an underlying essence. One study compared gender with 19 other social categories, such as race, sexual orientation, class, and religion, and found that gender was thought of as the most immutable, discrete, natural, necessary, and stable of all the categories.[1] It does not require much education to think of gender in these essence-like terms—even children as young as four rely heavily on the category of gender to make predictions about human behaviors.[2] Children recognize that a person's gender is largely fixed, and believe that even a girl raised only with boys and men would still retain feminine preferences. As a 10-year-old boy put it, when asked whether such a girl would prefer playing with a tea set or a toy truck, "because usually since she has a girl brain, she'd like to play with a tea set."[3] Young children actually have even more inflexible ideas about gender than do older children, and they adhere to these gender stereotypes so rigidly that they will sometimes insist that, say, nurses are women and doctors are men, even when they are shown photographs of female doctors and male nurses.[4] At a young age, children think of gender almost entirely as though it came from an essence.

It may seem curious that people, and especially young children, think about gender as coming from essences, because isn't gender just the product of how you were brought up? For example, one of the most acknowledged differences between boys and girls—that girls wear pink and boys wear blue—is not the result of any kind of deep biological connection. It's a recent cultural convention. Back in 1890, when American mothers read *The Ladies Home Journal*, they were provided with the clear instructions that they should use "blue for girls and pink for boys, when a color is wished."[5] Obviously much has changed since 1890, and this example reveals how norms and conventions often shape our perceptions of gender. There are also stark differences in gender norms across cultures, especially relating to gender equality. For example, only about 3 percent of elected officials in Arab nations are women, compared with about 45 percent in the Swedish parliament. One study asked both men and women from more than a dozen countries questions about sex role ideologies. In the most gender-egalitarian countries in the sample (the Netherlands, Finland, and Germany), the vast majority of both men and women agreed with items such as "Marriage should not interfere with a woman's career any more than it does with a man's," or "Women should be allowed the same sexual freedom as men," and both men and women usually disagreed with items such as "For the good of the family, a wife should have sexual relations with her husband whether she wants to or not," or "A man's job is too important for him to get bogged down with household chores." In sharp contrast, in the least gender-egalitarian countries in the study (Nigeria and Pakistan), the majority of both men and women disagreed with the items arguing for gender equality, and agreed with the items about women being subservient to men.[6] These cultural differences are the result of different societal gender norms (which have really changed only quite recently), and are not the product of more gender-egalitarian genes in northern Europe.[7]

The strong role that your upbringing has on gender norms is indisputable, and virtually all gender differences are influenced strongly by cultural practices, but this has led some to take the argument to the extreme, and to suggest that the *only* reason that women act feminine and men act masculine is because of their upbringing. Some parents, with the best of intentions, have tried to raise their children without any gender at all. The most radical perspective in this regard has been offered by some gender feminists who argue that the *only* difference between men and women is in their genitalia.[8] By this account, thinking of gender as emerging from an underlying essence would be pure folly—different gender norms would only arise from the arbitrary ways that society portrays men and women unequally. But, as we'll see both in this case, and throughout this book, explaining human development can never be adequately accounted for by looking either solely at arguments for nurture or solely at arguments for nature. The two are always interacting, and biological and psychological phenomena are the product of those interactions.

There is a reason that gender is the social category most rooted in people's thoughts about essences; biology and gender cannot be so easily separated. For the most part (with important exceptions to be discussed later), boys seem to be just born boys, and girls born girls. Take the unfortunate case of the identical twin boys Bruce and Brian Reimer, who were born in Winnipeg, Manitoba, in 1965. At the age of 8 months, they were taken to the hospital for routine circumcisions. Due to a tragic error, the electrocautery machine used to cut the foreskin burned Bruce's penis completely through "like a piece of charcoal." The famed gender psychologist John Money convinced Bruce's parents to approve an orchidectomy—at 22 months of age, Bruce's testicles were removed. The plan was that his parents would raise him as a girl and never let him know otherwise. Money believed that gender identity was nothing but the product of one's socialization, a common notion at the time. A girl could be made simply by raising her as a girl. Bruce's distraught parents saw no other choice, so Bruce was given the new name Brenda, and from that point on she officially became a girl. Together with her identical twin brother, Brian, who retained his genitals and male upbringing, Brenda became part of one of the most famous, and tragic, real-life scientific studies. Even though everything happened because of a horrific accident, Brenda's life story had all the trappings of a carefully designed scientific experiment. Brian, sharing the identical genes and family environment as Brenda, was the perfect control group. The selection of who would be raised as a girl and who would be raised as a boy was essentially random—determined solely by which boy was unfortunate enough to be subjected to the malfunctioning electrocautery machine. Any differences in their adult gender identity and behaviors, then, could be confidently attributed to the different upbringings that they received.

Brenda's parents and doctors were fully committed to the plan of raising her as a girl, and she was given dresses to wear and dolls to play with. No one at Brenda's schools was ever informed that she was anything other than a girl, and with her long curls and delicate features, there was nothing about her physical appearance that gave her away. But from the beginning, Brenda resisted all things feminine. Her brother, Brian, said in retrospect, "I recognized Brenda as my sister, but she never, ever acted the part." Indeed, Brenda seemed to fit the stereotype of a masculine boy about as well as anyone could: she ignored the dolls she was given and instead played army with her brother, she declared that she wanted to be a garbageman when she grew up, she regularly beat up the other boys in the neighborhood, and when her parents gave her a sewing machine as a present, the only thing she ever did with it was take it apart with a screwdriver.

As Brenda grew older, she fought hard against the notion that she was becoming a woman. When she was 12 years old, her father gave her the estrogen pills that Dr. Money prescribed and told her the pills would "make you wear a bra." Brenda threw a fit, shouting "I don't wanna wear a bra!" She resisted taking the pills, often flushing them down the toilet, until her parents forced her to take the pills in their presence. Brenda then became greatly disturbed when she started to develop breasts and hips, so she took to binge eating, consuming several ice cream cones each day, to put on some weight to conceal them. Whenever the discussion came up of Brenda receiving some future vagina reconstruction surgery so the doctors could, as they told her, give her a "baby hole," it would reliably set her off in an explosive panic; she absolutely hated the idea of developing a woman's body. And Brenda's interests remained starkly male-typical. She became the first girl in her vocational school's history to take a class in appliance repair, and she told her clinician about her detailed plans for building model gas airplanes and go-carts with CB radios. By then, she refused to wear anything feminine, and instead dressed in a denim jacket, torn corduroys, and construction boots. Her appearance and behavior led her classmates to nickname her "Cavewoman," and she was cruelly ostracized.

By age 14, Brenda had become so psychologically distressed that her parents and doctors worried she was becoming suicidal, and at this point her father took her out for ice cream to tell Brenda the truth. Brenda's feeling was one of instant relief—her deep feelings of alienation suddenly made sense. She immediately wanted to become a boy again, and she changed her name to David, to reflect how she had conquered this giant of her past. David's estrogen regimen was halted, and he later received penile reconstruction surgery. At the age of 25, he married a woman with three children from a previous relationship. A few years later, upon learning that other children were still receiving sex reassignments, in part because his own case had been described as a success by John Money, David decided he needed to go public to prevent other children from suffering the same fate that he did. He told his gripping story to John Colapinto, a reporter from *Rolling Stone Magazine*, who published an engrossing book in 2000 that described David's ordeal, titled *As Nature Made Him: The Boy Who Was Raised as a Girl*.

David's life ended tragically. His brother Brian developed schizophrenia, and in 2002, he died from an overdose of antidepressants.

On May 4, 2004, at the age of 38, two days after his wife had told him she wanted to separate, David drove to a grocery store parking lot and committed suicide by shooting himself in the head with a sawed-off shotgun. At the time, David was also struggling with unemployment, financial problems, recurring bouts of depression, and a troubled marriage, but it's hard to imagine that the horrific gender experiment he and his brother went through didn't also contribute to their tragic endings.[9]

The failure of David's experiment is echoed in a more expansive and systematic study that explored the experiences of 16 genetic males who were born between 1993 and 2000 with a devastating development defect called cloacal exstrophy. Cloacal exstrophy involves substantial malformations of the entire pelvic area that impair walking ability and digestion, can cause incontinence, and are associated with severely underdeveloped genitals. The parents of all 16 males were advised to have their children receive orchidectomies and vulva reconstructions shortly after birth, and to then raise them as girls without ever telling anyone about their child's sex. Fourteen of the families agreed to this (the other two raised their children as boys, with their testicles intact). Follow-up interviews with the families when the children were between 5 and 16 years old revealed that 8 of the 14 children raised as girls spontaneously identified themselves as boys, 5 identified as girls, and 1 vehemently refused to discuss anything about gender identity. All 14 children showed moderate to marked male-typical behaviors, which were absent among the biological sisters that 12 of them had, and all of them showed strong male-typical sexual behaviors and attitudes. Only 1 of the 14 children ever played with dolls or played house. It's worth noting that none of the 14 subjects were exposed to postnatal male-typical hormones, so their biological development was actually significantly less male-typical than those who retained their testicles.[10]

This study, together with the experiment that David endured, reveals that gender identity cannot just be assigned by fiat—there is an essence that underlies it. David's female upbringing could no better conceal his underlying male essence than Harry Potter's Muggle-rearing could repress his wizard essence. David's irrepressible masculinity was his nature.

The evidence for a biological essence underlying gender becomes even clearer when we look beyond humans. We have widely shared understandings about some boy-typical behaviors, such as rough-and-tumble play, playing with cars and trucks, and pretend play involving heroic characters like superheroes or soldiers, in comparison to girl-typical behaviors, such as playing with dolls, playing dress-up, and pretend play with family roles.[11] But what kinds of activities would be sex-stereotyped for monkeys, for example? Does the play of young monkeys parallel those of children? In an effort to answer these questions, researchers left pairs of human toys, a stuffed plush toy, such as a Scooby-Doo or a Winnie-the-Pooh, together with a toy vehicle, such as a dump truck or a bulldozer, in an enclosed area where 135 rhesus monkeys lived. The researchers observed the monkeys and noted their interaction with the pairs of toys. Interestingly, the male monkeys were several times more likely to play with the dump truck than they were with the Scooby-Doo doll, whereas the female monkeys showed no clear preference between the toys.[12] This mirrors the data from children, in which boys' play is more rigidly sex-typed than is the play of girls. These monkey data can't easily be accounted for by the notion that male monkeys are socialized to play with different toys than girls—none of the monkeys had ever seen anything like these toys before. And whereas playing with Scooby-Doo dolls might be seen to stimulate a maternal nurturing instinct toward infants among the girl monkeys—indeed, female wild chimpanzees, but not males, often walk around cradling a stick as though it was a baby[13]—the dump trucks remain more of a puzzle because wheeled vehicles play no part in monkeys' lives in any way.

The researchers proposed that the male monkeys' innate preferences for high activity rough-and-tumble play lead them to prefer toys with wheels. Even without knowing the precise biological mechanisms that attract male monkeys to trucks, this is more evidence that a biological essence plays a key role for some kinds of gender differences.

Although gender is the most essentialized social category that there is, even its boundaries are not set in stone. The simple division of people into two mutually exclusive genders becomes problematic when looking at intersex individuals, who are born with atypical sexual characteristics that prevent them from being distinctly identified as either male or female. But the limits of the biological determinism of gender are most apparent when looking at transgender people whose gender identity is at odds with both their biology and their socialization. Transgender people demonstrate that biology is not destiny. The gender you identify with can be quite contrary to the sex specified in your chromosomes—in many cases, even preschoolers may vehemently insist that their gender is different from what their parents have told them.[14]

But the responses that others often have to transgender people demonstrate the substantial cost of not fitting into conventional essence-based categories. In recent decades, it has become taboo for people to openly voice any prejudices toward many historically discriminated-against ethnic groups—the ethnic jokes that were a reliable staple in the 20th century have largely vanished from polite company, although such prejudices still remain widespread, as is evident when they are assessed with more disguised measures.[15] However, transgender people remain the target of some of the strongest prejudices that people will say out loud, and are frequently the victims of adverse hiring practices, police harassment, and violence.[16] For example, in a case that received much public attention, when the city manager of Largo, Florida, Susan Stanton (formerly Steven), confirmed to the *St. Petersburg Times* in 2007 that she was transgender, and was pursuing sex reassignment, her contract was promptly terminated, despite that she had received positive reviews about her performance until this news went public.[17] Stanton's case is not an exception—a recent survey on transgender prejudice found that almost 47 percent reported having been either fired, not hired, or denied a promotion because of being transgender. Likewise, 61 percent of those who expressed a transgender identity at school reported some kind of harassment, assault, or expulsion. And the discrimination that transgender people face is not limited to the public sphere, as 57 percent report some significant rejection by their own families. This discrimination has tragic consequences: the rate of attempted suicide among transgender people is an order of magnitude higher than the general population.[18]

The very notion of transgender people seems to be bothersome to many—in the words of the transgender author Kate Bornstein, "our mere presence is often enough to make people sick"[19]—and the prejudice and discrimination that transgender people frequently receive makes more sense when we consider it from the perspective of essences. When people expect the world to conform to its imagined underlying essences, it can be disturbing to encounter someone who seems to be acting in ways that do not fit with biologically assigned categories. Gender is thought to be homogenous and discrete—there are parts of identity that all men are perceived to have in common that are distinct from those of all women. This essence is thought of as natural and the ultimate cause of your gender identity. Transgender people are at odds with these perceptions, and this can be too unsettling for many people to accept. It is encouraging, however, that acceptance of transgender people has been rapidly improving, and their plight has received more attention in the popular media, for example, through such programs as *Transparent* and *Orange Is the New Black*, and through the substantial media coverage on Caitlyn (formerly Bruce) Jenner's coming out.

## Genetic Essentialism and Beliefs about Gender Differences

Although most people think about gender as having an underlying essence, they differ in the degree that they do so. Does it matter how much people think of gender in essence-based ways? Do people who think about gender as being grounded more in an essence think about men and women differently from those who essentialize less? For a rather extreme example, consider some gender-essentializing comments that have been made by the right-wing shock jock Rush Limbaugh. Responding to a CBS segment that discussed how women metabolize some medications differently from men, he went on a rant complaining about how feminists have pretended that gender differences didn't really exist: They would "say that there weren't any differences between men and women. . . . We could turn women into tough guys if we raise them the same way. What do you think the effort to have women take over the corporate world was? . . . They started replacing men, and now where are men, and where's our culture? It's been chickified in many, many places."[20] Limbaugh clearly takes the side that men and women are fundamentally different, and any attempts to reduce the distinctions between gender roles have been the source of nothing but problems. But alongside these gender-essentializing statements, Limbaugh has had a lot to say about women more generally. As a shock jock, he makes his living by saying startling and offensive things, but perhaps his most prolific way of offending people has been his frequent misogynist and sexist comments. For example, Limbaugh has called single mothers "semen receptacles," said that women shouldn't be allowed on juries when the accused is a stud, and his response to Sandra Fluke's argument for government-subsidized contraception was to call her a slut and a prostitute.[21] Is it just a coincidence that Limbaugh regularly makes both gender-essentializing comments and sexist ones?

A number of studies have explored what happens when people

think about gender in more essence-based ways. For example, one study asked German university students how much they agreed with the idea that genes determined various aspects of life by presenting them with questions such as "I think that differences between men and women in behavior and personality are largely determined by genetic predisposition." These same students then completed a measure of sexism that assessed how much people deny that women are discriminated against; a sample item is "Over the past few years, the government and news media have been showing more concern about the treatment of women than is warranted by women's actual experiences." The more that the students believed that genes determine people's lives, the more likely they were to show sexist attitudes.[22] Other studies have found that men who scored high on measures of sexism were especially likely to hold biological theories of sex differences. Convincing themselves that the sex differences are grounded in genes seems to reassure these men that women could never dislodge them from their privileged place.[23] Likewise, researchers in another study asked American female university students how much they agreed with items such as "Gender is more directly linked to biology than to the way a person is socialized," as well as to indicate how many stereotypically feminine traits (such as being soft-spoken, yielding, sensitive, compassionate) characterized themselves. The more these female students tied gender to biology instead of socialization, the more they identified with the stereotypically feminine traits.[24] Hence, those people who believe that gender has more of a biological basis are also more likely to believe that men and women really are different, and that we shouldn't be so concerned with fairer treatment of women.

We can see a related way that discussions of biological differences between the sexes can affect how we think about men and women. Consider what happened to one of the most powerful men in the United States. At a conference on diversifying science in January 2005, the former secretary of the treasury of the United States, and, at the time, president of Harvard, Lawrence Summers, offered a range of hypotheses for why women have been underrepresented in tenured positions at top universities in science and engineering. One of these hypotheses, which Summers called an "unfortunate truth," was that "my best guess, to provoke you . . . that in the special case of science and engineering, there are issues of intrinsic aptitude" between women and men.[25] He went on to say that "the human mind has a tendency to grab to the socialization hypothesis when you can see it, and it often turns out not to be true." A key reason why there are so few women in science and engineering, he argued, was not the product of sexual discrimination or of different socialization experiences. Rather, women were less likely than men to be born with the high level of math and science aptitude necessary to become world leaders in those fields.

In the rarefied air of the egalitarian and liberal environment of academia, these words went off like a nuclear chain reaction. They immediately led to heated accusations of sexism and charges of poor scholarship, and two months later, the Harvard Faculty of Arts and Sciences passed a motion of a lack of confidence in the leadership of Summers. This furor surely contributed to Summers's announcement to resign a year later.

But, you might ask, as several of Summers's defenders have, what is wrong with raising an interesting and legitimate scientific question? Are there intrinsic differences between the sexes that lead to the pronounced gender disparity in world-leading scientists and engineers? Wasn't Summers merely stating a hypothesis that could be investigated? If he was misguided with his essence-based musings, why not let researchers collect the data to prove him wrong? Because if we're serious about trying to understand what leads to this gender disparity in the sciences, shouldn't we be looking into *all* of the possible relevant factors, regardless of how provocative they are? This would be the best way to understand and potentially address the problem, and it is what we certainly would be doing if we were investigating a less politically sensitive question. Surely, no one would lose his or her job when suggesting that we should investigate whether intrinsic aptitudes could account for, say, the differing foraging behaviors of male and female fruit flies.

But, unlike fruit flies, studying humans can have far-ranging consequences for how they behave. The notion that intrinsic sex differences in aptitude can explain why there are relatively few female physicists resonates deeply with people's essentialist biases. People believe that men and women differ in their essences, so it makes sense to first look to those essences to account for the sex differences in scientific accomplishments. Although it's theoretically possible that any given difference between the sexes (such as scientific excellence) could have biological roots, the evidence that sex differences in math and science have a biological basis is far too equivocal at this point to support the confidence with which Summers made these claims.[26] But the bigger problem lies in our flawed thinking about genes. If people think there's an essence behind the sex difference in math and science, then they will likely think that the sex difference is natural and immutable. They very well might conclude that nothing can be done and resign themselves to a male-dominated world of math and science.

Actually, there's an even worse possible result of encountering such an argument. Consider the following study by the social psychologists Claude Steele and Joshua Aronson.[27] They gave both White and Black Stanford University students an extremely difficult test: namely, items from the Graduate Record Examination Verbal Test, of which, on average, only 30 percent of students can answer correctly. In the experiment, the students took the test under one of two different situations. Half of the students were just given the test as is. And after controlling for participants' own verbal SAT scores, the results showed that there was no difference between the performance of the White and Black students—they did equally well. The other half of the students took the exact same test, with the only variation being that they were asked to specify their race just prior to taking the test. For these students, there was a pronounced difference in performance between the White and Black students: the White students got about twice as many of the questions correct as the Black students did. What is going on here? How can this very subtle question of merely asking someone his or her race affect the performance of the students so much?

It was in this study that Steele and Aronson first discovered the social psychological phenomenon of stereotype threat. As social creatures, humans are aware of how they are viewed by others. We belong to groups, and people hold stereotypes about those groups, and we are all aware of those stereotypes. Now you might not believe in the stereotypes yourself, but you're still aware of them, and what Steele's and Aronson's research shows is that regardless of our beliefs about stereotypes, our awareness of the mere existence of them can affect our performance. If you asked Americans to list what they believed were the stereotypes that others held about Black Americans, one of them would be that Black Americans aren't as intelligent as White Americans—a stereotype that we'll return to, and will debunk, in Chapter 7. Even those Black Stanford students who have refuted it throughout their lives with their academic success are aware of this stereotype. And when they were asked to specify their race beforehand, the stereotype of less intelligent Black Americans was brought to the forefront of their minds. Stereotype threat is the anxiety that people feel when they are in a situation where they have the potential to confirm a negative stereotype about their group. Once this anxiety has been prompted, a cascade of psychological mechanisms are engaged, which conspire to leave the stereotyped person choking under pressure. Under stereotype threat, people start to feel stressed, they start monitoring their performance, they make efforts to suppress the negative thoughts that they're experiencing, they try to remind themselves of ways that they are unlike the stereotypes, their ability to focus on the task decreases, and they can become disengaged from the task itself.[28] In effect, they start acting in ways that confirm the stereotypes. The study of stereotype threat is one of the most popular and well-researched topics in social psychology, and it has been found across virtually all groups that bear a negative stereotype: for example, the memory of old people will get worse when they are reminded that they are old, the performance of White American athletes will suffer when they're reminded they're White, and women will get lower grades on math tests when they're reminded that they're women.[29]

My grad student Ilan Dar-Nimrod and I thought that perhaps the reason stereotype threat is so pernicious is that it often comes with the sense that the stereotype is grounded in an essence, and thus it would seem inescapable. Just as one's essence is thought of as immutable and underlying the identity of all who share that essence, then so should be the stereotypes that are associated with it. If women are worse than men in math because of the essence of their gender, then no woman would be able to escape the math stereotype because they all share the essence that makes them a woman in the first place.

To test this idea, we conducted the following experiment.[30] We had female Canadian university students take part in a study that was purportedly exploring factors that influence performance on the Graduate Record Examination—the exam that all prospective students need to take to be admitted to graduate school in several different disciplines. The actual test consists of a number of tests of extremely difficult math and verbal sections. The first part of the test was one of the math tests—this served as a measure of the student's baseline math performance. Next, the student took a verbal test, which included a reading comprehension task—this section contained the crux of our experiment. Each student read what appeared to be a newspaper article that was reporting on some scientific research, and was then asked some questions to assess comprehension of the article. Actually, none of the newspaper articles in our study was real; we had completely fabricated them in order to manipulate how our participants were thinking about gender and math. One group of students was randomly assigned to a "math genes condition." This group read an article that described how scientists had recently discovered the existence of some "math genes," which were located on the Y chromosome, which meant that only men had them. The article went on to say that men perform better than women on math tests by about 5 percent, and that these math genes account for the difference. Another group of students was assigned to a "math socialization condition." These students read an article that described how scientists had recently discovered that teachers teach math differently to young boys than girls, and that this difference in teaching styles explains why men outperform women on math tests by about 5 percent. And a third group of students was assigned to a "no sex difference in math condition." This group read an article that described how scientists had recently done an extensive survey of the entire published literature, and had concluded that there were actually no sex differences in math at all. I should emphasize that all of these articles were completely made up, and there was no truth to any of the claims in them—at the end of the study, we told all the students about how all the information in the articles was fabricated, and we then explained the true purpose of our study.

After the students finished reading this article and the other questions in the verbal section of the test, they then took a second math test. When we compared how well people did who read that there were no sex differences in math compared with those who read about the existence of male math genes, we found a clear replication of past research on stereotype threat. When students were struggling on an extremely difficult math test, and they were thinking about the existence of male math genes, the stereotype about women and math seemed inescapable. They ended up choking on the second math test. In contrast, those students who read the "math socialization" article performed at the same level as those who had read that there were no sex differences in math. They showed no stereotype threat, even though students in both the math genes condition and the math socialization condition were informed (falsely) that men outperform women by 5 percent in math. Thinking that the stereotype was caused by people's experiences made it seem external to their essences—if it wasn't really part of them, it could be escaped. But if the stereotype was grounded in your genes, and was thus part of your essence? Well, you can't hide from your genes.

These findings appear to vindicate the critics of Lawrence Summers, in that they demonstrate that some ideas can produce undesirable effects. By merely airing a genetic hypothesis about why women are underrepresented in science and engineering, Summers may well have contributed to making the gender gap wider. But what should we do with this knowledge? On the one hand, these results show that we could potentially improve women's math performance by teaching everyone that sex differences are the result of socialization and not genes. But remember that in our study we made all of this information up. There is much evidence that females are underrepresented in the highest echelons of math and science achievement, and the actual causes of these sex differences continue to be vigorously debated.[31] Although declaring that any sex differences in performance are the result of socialization may potentially work toward reducing any sex differences in the future, this is not how science operates. We don't know how the results of our experiments will turn out in advance, and there is no rule that nature only operates according to politically correct principles. It remains entirely possible that there are biological differences between the sexes that are relevant for math performance, although this hasn't confidently been shown yet.

## Genes and Sexual Orientation

People like to have sex, and most people like to have sex exclusively with people of a particular gender. But the simple question of which gender people would like to have sex with remains a remarkably politically contentious one. Many people, typically social conservatives, find the idea of homosexual relationships to be deeply bothersome and sinful. Their anger and disgust can be strong enough, in fact, that issues of gay and lesbian rights have sometimes been used to motivate social conservatives to get out and vote, as was done in the United States with the issue of same-sex marriage in 2004.[32] Two ongoing wars, dramatic changes to tax policies, and the privatization of social security, were not enough to motivate some to the voting booth—for many, only the gay and lesbian rights issue upset them enough to vote. By some estimates, George W. Bush would not have been elected to a second term without the issue of same-sex marriage being on the ballot in the state of Ohio.[33] And it's not just in the United States that homophobia remains widespread; Vladimir Putin sparked international outcry by criminalizing "gay propaganda" in advance of the 2014 Sochi Olympics, and there are still several countries in Africa and the Middle East where sodomy is an offense punishable by death. Why do so many social conservatives care so deeply about the issue of other people's sex lives?

Much of our attitudes toward gay people pivots on what we think *causes* our sexual orientation. Several studies have revealed that the more people view homosexuality as something that you are born with, the more tolerant they are of it.[34] For example, the more you agree with the item "Whether or not a man is homosexual or heterosexual is pretty much set early on in childhood" the less likely you are to agree with items such as "I think male homosexuals are disgusting."[35] Not surprisingly, then, those who are most deeply offended by the existence of gay people, such as conservative fundamentalist Christians, tend to think that homosexuality does not have a biological basis. Rather, they are more likely to believe that people choose to be gay. Many social conservatives are antigay because they view heterosexuality as the "natural" way—the way God intended—which means that homosexuality must be seen as unnatural, and, as the naturalistic fallacy thus dictates, a moral failing. By this reckoning, if someone is gay, it is because the choices that person has made are seen as sinful, in the same way that choices to engage in criminal behavior are seen as sinful. But as the social psychologist Jesse Bering notes, this argument really only makes sense for someone who has bisexual preferences.[36] If someone only has heterosexual attractions, why would that person ever "choose" to be gay? It's hard to see a behavior as a choice if one doesn't have a temptation to act in the first place.

But of course, many people don't see sexual orientation as a choice in their lives. Rather, many see it as a core part of their essence— it reflects who they fundamentally are and always have been. For example, Lance Bass, the former singer from the boy band 'N Sync, opens his autobiography with the following statement: "I've known I was different ever since I was five years old. For one thing, I had what I guess you could call innocent crushes on boys."[37] This kind of essence-based accounting for the origins of homosexual attraction is very much the norm, in particular, for gay men. As with essences, many gay people think of their sexual orientation as immutable—it's always been a deep and fundamental part of them, it's in their nature, and they share this essence with other gay people, and this shared essence makes them distinct from heterosexuals.

Although the notion that one's essence is, and always has been, gay, is commonly experienced among gay men, it is a view that is less typically embraced among lesbians, although it's still not uncommon among them. In general, women seem more flexible than men with regards to their sexuality;[38] for example, both gay and straight men are less likely than lesbians and straight women to see their sexual orientation as a matter of choice.[39] Gay men thus do seem to be more likely to be "born this way" compared with lesbians, and because of this, the majority of research on the causes of homosexuality have been conducted with men.

So if the theories that people have regarding the causes of sexual orientation are associated with the strength of their antigay prejudice, could it be possible to reduce prejudice by changing people's theories about the causes of homosexuality? If a biological basis of homosexuality became the most widespread theory about the origins of sexual orientation, would homophobia become a relic of the past?

First, it's important to remember that there's a critical distinction between causal evidence and correlational evidence, and the majority of the research that has linked positive attitudes toward gay people to the notion that people are born with their sexual orientation has been correlational research. And because correlations do not imply causality, it's possible that the association between believing that sexual orientation is chosen and in holding antigay prejudice is due to some other factor (such as people's disgust sensitivity).[40] But there is also evidence that people's theories of the origins of sexual orientation *cause* people's antigay prejudice—this means that learning a new theory for where sexual orientation comes from can change whether or not a person feels homophobic. For example, one study had Spanish university students read an essay that either referred to scientific research showing that people were gay because they were born that way, or about research showing that people were gay because they were *raised* that way. Later, the students were asked how much they supported gay rights and same-sex marriage. Those students who had read that being gay was the result of how you were raised were more opposed to gay rights and same-sex marriage than those who read about gay people being born that way. For example, those who read that sexual orientation was the result of how you were raised were more likely to agree with the statements "If a child is adopted by a same-sex couple, he/she will surely have psychological problems in the future," and "I think it is a social error to legalize marriage between people of the same sex."[41]

A version of the experiment above has also played out in real life. In 1993, the geneticist Dean Hamer published evidence that a genetic marker was linked to male homosexuality. His team was studying a sample of gay men, and the researchers asked the participants about the sexual orientation of their relatives. The responses indicated that only certain relatives of the participants were more likely than average to be gay: the gay men reported that more of their maternal, not paternal, uncles were gay, and more of their cousins who were sons of maternal aunts were gay than were their other cousins. This pattern suggests that men's sexual orientation might be associated with the X chromosome, which men receive exclusively from their mothers, and share with their brothers, maternal uncles, and male cousins of their maternal aunts. And, indeed, by analyzing the DNA of 40 pairs of gay brothers, Hamer's team identified a genetic marker on the X chromosome of these men, Xq28 (a region that contains several genes), that looked like it might possibly predict sexual orientation. Specifically, 33 of the 40 pairs of gay brothers had similar genetic variants along Xq28, whereas only 20 pairs would have been expected to share these by chance. The strength of this association was rather modest, there were no specific genes identified, and the authors of the paper were quite careful in highlighting just how tentative and ambiguous their findings were. But this was still the first published evidence of a specific genetic marker associated with homosexuality. So how did people respond to this weak and tentative scientific finding?

As Dean Hamer put it, "Rarely before have so many reacted so loudly to so little."[42] This one simple research finding led to a seismic overhaul of the political landscape. Almost every major newspaper in the United States reported the results of this study on the front page,[43] and a flood of newspaper columns, magazine articles, and talk shows discussed the implications of these findings. Not surprisingly, these discussions frequently bore the distinct marks of switch-thinking and essentialist biases. Typically, the discussions centered around what this discovery of "the gay gene" meant, even though no specific gene was found, and the authors never used this term. And although the actual finding only indicated the *possible* existence of some genes that have an influence on male sexual orientation, the media seemed to conclude that scientists had discovered an essence underlying homosexuality. The most common argument in the media coverage went like this: if homosexuality was genetic, then it was immutable and natural, and should be accepted. The *Boston Globe* ran an editorial that argued that this discovery of a genetic basis of homosexuality would go "toward undermining the narrowly gauged argument that homosexual behavior is misconduct . . . [and] could ease the struggle to secure equal protection."[44] Gay rights activists, such as Greg King and Richard Green, were quoted several times in the national media for saying that the "evidence could be liberating," that this was a "landmark study . . . [for] civil rights" and "immutability" could lead "to an overturning of discrimination laws."[45] And these essentialist predictions may have been quite prescient: prior to Hamer's research, in 1985, only 20 percent of Americans thought that homosexuality was something "that people are born with," but by 2012 this had doubled to 41 percent.[46] The growing acceptance of homosexuality may well be tied to how our essentialist biases make sense of the increasing evidence of genetic influences on sexual orientation.[47] If so, this would be a case where thinking about underlying essences can lead to positive outcomes for the gay rights movement.

Even many who are vehemently antigay find this kind of genetic evidence to be an unassailable argument in favor of gay rights, so if people are to maintain their homophobia, they need some way to undermine this evidence. For example, the controversial Christian radio host Bryan Fischer argued against the genetics of homosexuality by questioning the validity of the heritability of sexual orientation from twin studies. He pointed out that for identical twins, "If one of them is gay and it's genetically caused, the other one ought to be gay one hundred percent of the time!"[48] Because this is clearly not the case—a gay identical twin shares the sexual orientation of his brother somewhere between 30 percent to 50 percent of the time—Fischer saw this as evidence to refute the argument that sexual orientation has a genetic component. Fischer's switch-thinking reasoning fits with the genetic essentialist bias of genes as *ultimate causes*—having a gene is akin to a diagnosis, so anyone with a "gay gene" would have to be gay "one hundred percent of the time." Hamer's search for genetic markers of homosexuality has thus been deeply moving to both critics and supporters of gay rights.

This is a relatively rare situation where people's stance on a sensitive political issue pivots around the results of what is ultimately a scientific question—are genes involved in sexual orientation?

## The Darwinian Paradox and Explanations for the Origins of Homosexuality

There is another aspect of "gay genes" that many find problematic: If gay genes exist, where could they have possibly come from? The Darwinian paradox refers to the problem with positing the existence of genes that predispose someone toward homosexuality because those who would possess them would be the least likely to reproduce and pass them on to the next generation. Indeed, research finds that the fertility of gay men is much lower than that of their straight counterparts—for example, a recent Italian study found that gay men have only about 20 percent the number of children as straight men[49]—a figure comparable to that found in other studies (although the difference in the fertility between lesbians and straight women is much smaller).[50] And behavioral genetics studies with twins show that, like almost everything else, sexual orientation has a substantial genetic basis; it's about 30 percent to 50 percent heritable, and there must be many different genes involved. Hence, on the surface, there really does seem to be a paradox, as the twin studies show that genes somehow contribute to homosexuality, yet this would mean that the genes of gay men would seem less likely to make it to the next generation. It's a similar problem that the Shaker religion has faced; the religion, which reached its zenith in 19th-century New England, banned sexual intercourse among its members, yet the most common way that people acquire their religions is by learning from their parents. Hence, the Shaker movement has dwindled steadily from its high-water mark of several thousand members to a grand total of three people in 2009.[51] So if genes influence sexual orientation, how come gay people haven't gone the way of the Shakers?

There is not a single obvious answer for how homosexuality could be common if there's a genetic basis, and the Darwinian paradox has spawned several distinct theories and lines of research.[52] And because the sexual orientation of men appears more stable across the lifespan than among women,[53] almost all of the research that has tried to make sense of the Darwinian paradox has been based on studies of gay men.

First, in considering how genes get passed down to future generations, it's important to recognize that your genes aren't just possessed by you—they're also possessed by your family members to various degrees. Your biological parents, siblings, and children each share on average 50 percent of your genes; your aunts, nephews, grandparents, and grandchildren each share about 25 percent of your genes, and your cousins each share about 12.5 percent of your genes. This means that your genes can appear in the next generation even if you don't have any children yourself. Take my own case as an example: setting aside my two children, my biological relatives (that I know of) who are one generation younger than me consist of my 4 nephews and nieces, and 31 first cousins once removed (children of my cousins). My extended family is quite fertile, and if you do the math for my genetic relatedness to all of my relatives' children, this means that about 294 percent of my genes appear in my relatives in the next generation outside of my immediate family, and this number doesn't include the many, many distant relatives who also share smaller portions of my genes. This latter portion wouldn't be insignificant either—according to 23andMe, where I've been genotyped, there are more than 1,000 people in the company's database who appear to be related to me, although the vast majority of these are so distantly related that we only share about a thousandth of our genes. What this means is that, regardless of my own reproductive success, my genes will still be passed down to future generations by my biological relatives. Genes that predispose someone toward homosexuality might thus be selected for if gay men have more fertile relatives than heterosexual men do.

One possible mechanism for this is that perhaps genes associated with male homosexuality are actually genes that increase your androphilia—that is, your attraction to men—so that female relatives of gay men, who possess these same genes, may be especially androphilic, and this would make them more likely to reproduce. And there is some evidence to suggest this is true: for example, one study found that the maternal female relatives of gay men had significantly more children than the maternal female relatives of straight men.[54] The genes of gay men might thus thrive because of their unusually fertile sisters.

A related way of explaining the Darwinian paradox was offered by the sociobiologist E. O. Wilson,[55] who proposed that "kin selection" could account for the persistence of genes associated with homosexuality. Kin selection refers to how the survival of our own genes is facilitated by any motivations to help our kin, who share our genes. The stronger our sense of love and devotion to our kin, the more resources and efforts we devote to helping our kin survive, and the greater the likelihood that our genes will make it to the next generation. It makes some intuitive sense that people are more motivated to help their relatives than they are to help nonrelated individuals. For example, the Carnegie Hero Fund Commission bestows medals on those who risk their lives to help save others, yet one of their requirements for bestowing medals is that the person who is saved cannot be an immediate family member of the rescuer. Although rescuing a family member may be commendable, it's not seen as heroic, but rather expected, according to the logic of kin selection. So perhaps gay men contribute to the survival of their kin's genes by being more likely to help out their nephews and nieces compared with straight men. Twenty-first-century Western societies probably aren't the best places to look for support for this hypothesis[56] because modern technology and social welfare systems have reduced the need to rely on one's uncles to meet survival needs. However, in at least one subsistence culture, in the Samoan Islands, there is solid evidence that a group of biological males known as the *fa'afafine* (who identify as a third-gender people who are attracted to men) do show more avuncular devotion to their nephews and nieces than do straight Samoan men by babysitting and contributing money to medical care and education.[57] If these results held with other subsistence populations around the world, the kin selection hypothesis could help explain our Darwinian paradox. The genes in question wouldn't just make men sexually attracted to other men; they would also make them more devoted to caring for their nephews and nieces.

Of course, the kin selection hypothesis would only help us if gay men had a lot of nephews and nieces—if there weren't many nephews or nieces around, then there would be fewer selective advantages to having genes that led you to invest more in your siblings' children. It turns out that gay men often do have a lot of nephews and nieces, but one of the biological mechanisms that underlie this is not genetic. Rather, it's a response to the mother's immune system. When women are pregnant with a son, they produce some antimale antibodies as an immune response to the male-specific antigens linked to the Y chromosome. This immune response builds with each successive male fetus, such that later-born sons are bathed in a womb full of these antibodies, which some have hypothesized may affect their later sexual preferences.[58] In fact, there is abundant evidence that men who have a lot of older brothers are more likely to be gay. Ray Blanchard, a psychiatrist, found that for every older brother a man has, he is 33 percent more likely to be gay, and this relation has been observed in several countries around the world. If you extrapolate from this evidence, then the rare man who has ten older brothers would have about a 50-50 chance of being gay. The effect is strong enough that about one-quarter of all gay men could potentially attribute their sexual orientation to their older brothers.[59]

And yet another way that genes associated with homosexuality might become common in subsequent generations is that the genes might predispose people toward temperaments that aren't directly tied to sexual orientation *per se*, but that early childhood experiences might then shape the sexual orientation of people with those temperaments. The "exotic-becomes-erotic theory," which was proposed by the social psychologist Daryl Bem, argues that what ultimately distinguishes gay men from straight men is their differing preferences for high-arousal activities as children, like rough-and-tumble play and sports.[60] Indeed, much evidence finds that gay men report being less likely than straight men to have engaged in much aggressive behavior as children, and they also report liking sports less.[61] Likewise, gay men report having had more female friends as a child than straight men do, and of feeling different and unfamiliar when around most other boys.[62] This "feeling of difference" is hypothesized to create a sense of discomfort and tension whenever they are around other boys. A similar feeling of difference and discomfort seems to exist when young heterosexual boys are around girls, but then, some time around puberty, most heterosexual boys go through a transformation where that discomfort and tension go from a sense that "girls are yucky" to a feeling of sexual yearning. Likewise, among boys who feel different from other boys in their dislike of male-typical activities, they will also come to interpret the discomfort and tension that they feel around other boys as feelings of sexual arousal when they go through puberty. Indeed, much psychological research has shown that people can reinterpret various extrinsic sources of arousal (such as elicitors of violence, fear, or excitement) as indicators of sexual arousal.[63] If this theory is correct (and there's supportive evidence for many of the hypothesized steps involved, but more definitive research is needed), it would show that the key genetic differences between gay and straight men influence their attraction to high-arousal activities, such as sports and rough-and-tumble play, and these would get translated into feelings of homosexual attraction for only some boys. Because many boys with these same genes would not reinterpret their feelings of tension as homosexual attraction, the genes remain common in subsequent generations.

It's worth noting that there have been many other theories of homosexuality that don't involve genes; these other theories don't have to deal with the Darwinian paradox, but they also don't seem to have had much impact on shaping political discourse. For example, a key psychoanalytic theory about homosexuality is that gay men tend to have had overcontrolling and smothering mothers who acted seductively toward them, and fathers who remained detached or hostile toward them.[64] Even though it's clearly beyond a person's control to change his or her parents or childhood, these psychoanalytic arguments did not lead to more tolerant attitudes toward gay men. It's only genetic arguments that have had such a dramatic impact on antigay prejudice because only genes point to an underlying essence for sexual orientation.

## The Downside of Gay Essences

But there is also a dark side to linking genes with sexual orientation. The link between essence-based thinking and attitudes toward gay people is not uniformly positive. The social psychologists Nick Haslam and Sheri Levy found that essence-based beliefs about homosexuality had three distinct clusters. The first cluster was a belief that homosexuality was biologically based and immutable, and the second cluster was a belief that homosexuality was universal across history and cultures. For both of these two clusters, believing in an underlying essence of homosexuality was associated with less anti-gay prejudice. The third cluster was a belief that homosexuality was discrete—that is, a belief that there are clear differences between gay and straight people. However, the more that people believed in this particular essentialist basis of homosexuality, the *more* prejudiced they were against gay people. Thinking about gay people as fundamentally different from heterosexuals tended to go hand in hand with homophobia, and it also led to another disturbing response. Thinking that genes influence people's sexual orientation led many people to propose that eugenic technologies could be used to prevent people with genes that predispose them toward homosexuality from ever being born.

In their reporting of Dean Hamer's original findings, the *Daily Mail* headlined the story "Abortion hope after 'gay genes' findings." As outrageous and offensive as this headline was, it reflected an argument made often in the media coverage of this research. Sir Ian McKellen, the founder of the gay rights group Stonewall, wrote that the dominant theme to emerge from this public discourse was "whether a mother should or should not have the right to abort her gay foetus. In other words, what a problem we gays cause our parents."[65] Two newspapers furthered this discussion by printing lists of famous people, such as Tennessee Williams, Virginia Woolf, Michelangelo, and Martina Navratilova, who they said would have been lost to humanity had the genetic technology for selective abortions for gay people been available earlier.[66] A former chief rabbi of the UK, Rabbi Jakobovits, wrote in a pair of letters to the *Times* that homosexuality was "a grave departure from the natural norm which we are charged to overcome like any other affliction, genetic or not," and argued that "the errant gene" should be "removed or repaired," to liberate the afflicted from their "disability."[67] Similar kinds of responses were identified in a study that interviewed a broad sample of American adults and asked their response to the question "If it were found that homosexuality is partly genetic, how do you think that information might be used to help or harm people?" The responses included such homophobic gems as "Well, as far as I'm concerned, it'd be good to correct whatever the problem is in the gene and have all straight people," and "I'll use the leprosy thing: They're going to put people on an island and separate them."[68] It's striking to see how many people so easily turn to futuristic genetic engineering technologies to "cure" homosexuality, whereas the regular news coverage of newly discovered "cancer genes" rarely sparks such enthusiasm for eugenic solutions.

The quite volatile reactions that people have had to discussions of genes and homosexuality make more sense when you consider them from the perspective of genetic essentialism. If genes are shown to be involved with sexual orientation, then our essentialist biases kick in. On the progay side, our biases suggest that sexual orientation is *immutable* and *natural*, and thus should be accepted, but on the anti-gay side they suggest that homosexuality is *discrete*, and it prompts discussions for eugenic solutions. To me, all of these reactions are irrational; evidence for biased thinking, rather than evidence for any reasoned weighing of scientific arguments. Whether or not genes are involved in sexual orientation says nothing about its stability or about its desirability. The hopes or fears associated with finding a "gay gene" reflect our switch-thinking biases—there is no simple genetic switch to be found here.

It's highly unlikely that any individual genes will ever be identified that are strongly predictive of sexual orientation. It's worth noting that not only were the effects that Hamer found in his study extremely small, nor did the study identify any specific genes, but also other labs have had a hard time replicating these findings.[69] In all likelihood, the genetic influence on sexual orientation is a lot like the genetic influence on other complex human traits. It will look much more like the genetics for height, with dozens, if not thousands, of genes that interact in complex ways with experiences, rather than like Harry Eastlack's fully penetrant monogenic condition of FOP.

Although some people, such as Lance Bass, may have always felt gay, this does not mean that they possess a unique gay genetic switch that distinguishes them from straight men. Rather, the web of forces that make some men gay, and the majority of others straight, are part of a complex interaction involving many genes, the potential involvement of epigenetic markers that guide their expression,[70] prenatal experiences in the womb, and the collective influences of life experiences along the path of development. Your sexuality is the product of the interplay among all of these factors. Genes play a key role in shaping our sexual preferences, but they most certainly do not do so in the deterministic manner that people attribute to essences.

Whether people are thinking about gender or sexual orientation, they're often imagining an underlying essence that makes things as they are. When these essentialist thoughts point to a natural basis of something that has traditionally been morally condemned, as in the case with homosexuality, people develop more favorable attitudes toward it. On the other hand, when we think about men having different essences than women, or gay people having different essences from straight people, it can fan the flames of our prejudices. In the next chapter we'll discuss how those same reactions to essences shape how we think about race and ancestry.

# 6.

# Race and Ancestry: How Our Genes Connect and Divide Us

**THE RITUAL HAS BEEN PERFORMED COUNT**-less times over the millennia, but perhaps never with quite the same implications. With a steady hand, and making a precise cut, the mohel carefully removed the foreskin and completed the circumcision in the apartment of Rabbi Baruch Oberlander, above the synagogue in Eastern Budapest. What made this particular circumcision special was the person on the receiving end of the scalpel. Csanád Szegedi, a 31-year-old father of two, and a member of the European Parliament, had recently committed to living his life as an Orthodox Jew. Szegedi is currently learning Hebrew, attends synagogue every Friday, eats kosher, has taken on the new name of Dovid, and is doing his best to follow the 613 commandments that govern the lives of devout Jews. So far he has only been able to conform to about 80 of those commandments. "I am trying," he says, "but it doesn't happen overnight."[1]

We might forgive Szegedi for taking some time to adjust to the life of an Orthodox Jew because he does have more to learn than most. After all, the seat that he occupies in Parliament was one that he was first elected to as a member of the Jobbik Party in 2009, an extreme right-wing and anti-Semitic party in Hungary. Szegedi wasn't just one of the party's rank and file, he was the vice president, and he also had been a founding member of the Hungarian Guard, a now-banned paramilitary group who wore black uniforms to evoke Hungary's wartime Fascist party. Szegedi even showed up to his first day of work in Parliament wearing his Guard uniform. He recently published a 316-page book titled *I Believe in Hungary's Resurrection*, which was filled with his anti-Semitic speeches and writings. But because of his new Orthodox Jewish identity, he asked bookstores to return several thousand copies of the book, which he dumped in an oil drum and set on fire. "With this I was able to cleanse my own past and bring that era to a close. The cover has my picture on it and it was very strange to see my face burn."[2]

Rarely does one witness such a dramatic about-face as this, and it follows that something awfully momentous must have occurred in Szegedi's life for him to undergo such an enormous transformation. And, indeed, recently it had: Szegedi discovered that he was a carrier of Jewish genes.

How a grown man, obsessed with anti-Semitic conspiracy theories, and raised by his biological family, could possibly not know about his own Jewish origins says a lot about the tortured relations with Jews that still persist in much of Eastern Europe. Szegedi's maternal grandmother, Magdolna Klein, was a survivor of Auschwitz. But until her death in 2014, she feared that the Holocaust would repeat itself, so she and her husband, who had also been in an internment camp, decided not to tell their daughter, Szegedi's mother, about their past. Magdolna only wore long-sleeved shirts, even in summer, to conceal the prisoner number that was tattooed on her arm. In 2010, a political rival, Zoltan Ambrus, met with Szegedi and showed him a copy of his grandmother's birth certificate that had mysteriously fallen into his hands. Believing that it was all a lie, Szegedi did nothing, but in Christmas 2011, he broached the topic with his grandmother. To his shock, she confirmed that the stepparents who had raised her were Jewish. This was obviously disturbing news for Szegedi, but as he said, "I calmed down, because it's only the stepparents—they are not blood relations of mine."[3] But the next Easter, when he saw his grandmother again, he pushed her for more information. It turned out that his grandmother's stepfather had actually been her mother's brother, who had taken over her care after her mother had died. This meant that Magdolna, and thus Szegedi himself, was genetically Jewish by descent.

At first Szegedi tried to keep things secret, but in June 2012, just as Szegedi's reelection campaign was taking place, a right-wing website published the birth certificates of Szegedi's grandparents. Moreover, the Jobbik Party had videotape of Szegedi allegedly trying to bribe Zoltan Ambrus to keep his Jewish history secret, and used this bribery as pretext to force him to resign from the party. But Szegedi still won his reelection and retained his seat in Parliament. The week after he resigned from the Jobbik Party, he reached out to the Rabbi Slomó Köves and began his accelerated path to becoming an Orthodox Jew. "I am just as Hungarian as until now," Szegedi stated, "but I have expanded my own identity with the Jewish identity."[4]

Szegedi's remarkable transformation lays bare an important truth about our identity. Much of who we think we are is based on what we think about our own biological roots. Szegedi's profound change of identity ultimately hinged on the evidence that his grandmother's biological parents were Jewish—the belief that she was raised by Jewish stepparents wasn't enough to persuade him. And it doesn't seem to matter to Szegedi that he hadn't been raised Jewish in any way—his parents were Christian, his mother had no idea of her Jewish roots, and he had no experiences, or knowledge, about the cultural traditions. His newfound identity stems from his belief that who he is as a person is ultimately determined by who his genetic ancestors were. This information about his biological roots led Szegedi to realize the unbearable hypocrisy of his political beliefs; his whole career had been predicated on attacking people whom he felt shared his very essence.

Szegedi is certainly one who believes in the power of essences. Discovering the truth about his own Jewish essence didn't just lead him to reject his anti-Semitic past. Rather, his essentialist thinking runs so deep that he seems to think having Jewish genes means that he needs to be as purely Jewish as possible; hence, his conversion to orthodoxy. But Szegedi's story, however bizarre and extreme it may be, shows that when it comes to our sense of identity, our essentialist biases once again lead us to view genes as ultimate causes.

A first step to understanding the source of your essence is to look at your parents. You may be able to recognize in them the shape of your nose, your sense of humor, or your curious preference for the way you organize your cutlery drawer. Your parents can seem like your own personal portrait of Dorian Gray, aging before your eyes, one generation ahead of you. Because most people are raised by their biological parents, they typically have a fairly clear sense of where their essence comes from. But the existential security that comes from knowing about one's biological roots may be elusive for those who have never met some members of their biological family. Not knowing one's biological ancestry can be accompanied by a profound sense of alienation, identity confusion, and "genealogical bewilderment."[5] This quest for a biological basis of identity is perhaps most poignantly evident in adoptees, who often struggle with a feeling that the adoption researcher Betty Jean Lifton calls "a hole in the center of their being."[6] Many adoptees feel stuck as they try to negotiate the gap between the identity that they've acquired through their life experiences and the sense of missing identity that comes from not knowing about their biological essence.[7]

People are often fascinated to learn more about their biological essences, and this quest inspires both adoptees and nonadoptees alike.[8] This fascination with essences is so widespread that, according to *Time Magazine*, genealogy has become the second most popular hobby in the United States after gardening, and is the second most visited category of websites after pornography.[9] But this quest to learn more about one's biological roots has recently been radically transformed by a new technology. In the past several years, approximately 40 direct-to-consumer genomics companies have emerged that provide people with an avenue to discover their biological ancestry by analyzing their genes.[10] In what would have recently seemed like a far-fetched science fiction novel, these companies can peer in and read the text that is written in the string of nucleotides in your cells. And they can decipher this script and inform you about the likely geographic origins of your forebears. Exploring one's genetic ancestry has rapidly become a popular pastime, and it is estimated that more than 3 million of these tests have been sold worldwide since 2000.[11]

## Searching Your Ancestry in Your Genes

How is it possible that our DNA can provide information about where our ancestors come from? To answer this question we first need to consider the early history of our species, *Homo sapiens sapiens*. The first anatomically modern humans emerged in the savannah of eastern Africa approximately 150,000 to 200,000 years ago, although the precise dates, as well as the dates of subsequent migrations, continue to be debated.[12] They spread out throughout Africa until about 50,000 to 60,000 years ago, and then, for some reasons that aren't completely clear, a relatively small group of humans decided to leave. That group crossed the Isthmus of Suez into the Middle East, and some of their descendants followed the coast around India, all the way through to Indonesia, Papua New Guinea, and then arrived in Australia about 45,000 to 60,000 years ago.[13] Other groups of humans continued to leave Africa, and some of them headed north through central Asia around 35,000 years ago. At this point, some early humans turned to the west and began the human settlement of Europe. Around the same time, another group headed east and populated northeastern Asia. Some of those living in Siberia then headed further east, crossing over a land bridge that spanned the Bering Sea, entering Alaska, and then took a sharp turn south, ultimately reaching the tip of South America approximately 10,000 years ago. Last, a group of adventurous Austronesians left New Guinea aboard outrigger canoes to populate the scattered islands of the South Pacific, not reaching the furthest outposts of Hawaii, New Zealand, and Easter Island until about 800 years ago.[14]

These massive migrations, which have resulted in the human colonization of all habitable regions of the globe, preceded any written records, and even the archaeological record is filled with gaps. How then do we know about the particular paths that were taken? It turns out the migrations of early humans are recorded with remarkable precision in our DNA.[15]

There are three distinct ways that genes reveal ancestry, and each of these ways is exploited by direct-to-consumer genomics companies. First, each of the cells in your body possesses some mitochondria (the power plants of your cells), and these contain a small circular genome that is separate from the DNA in your chromosomes. Because spermatozoa jettison their own mitochondria, the mitochondria in your cells come exclusively from the egg cell provided by your mother. This is a key factor, as the vast majority of your chromosomal DNA combines between your mother and father, shuffling their distinctive genetic contributions to you, and thus concealing which genetic variants came from which parent. In contrast, because your mitochondrial DNA comes solely from your mother, it never gets shuffled. You inherit it wholesale—the mitochondria in your cells are largely identical to the mitochondria in your mother's cells, which are the same as the mitochondria in her mother's cells, and so on back through to time immemorial. However, even though mitochondrial DNA does not get shuffled with each generation, every now and then there is a copying error that leaves one with a mutation in the mitochondrial DNA that is slightly different from that of his or her mother's. For the most part, these mutations have no biological consequence in our bodies, but it is because of these copying errors—which are extremely rare and thus are straightforward to detect—that we can learn about our ancestry from our DNA.

Let's say a woman, we'll call her Nefertiti, is born with a de novo (new) mutation in her mitochondrial DNA. That mutation is an indelible marker that will then be inherited by all of Nefertiti's future descendants. You can think of that mutation as akin to a stamp in a passport. Each of Nefertiti's descendants will possess that stamp, and thus any people who have that same stamp today must ultimately be descendants of Nefertiti. Likewise, let's assume that Nefertiti's great-great-great-granddaughter, we'll call her Cleopatra, is born with another de novo mutation in her mitochondrial DNA. This is a new stamp in the passport, and all descendants of Cleopatra share this same stamp. In contrast, Cleopatra's sister, we'll call her Cornelia, didn't possess the same mutation as Cleopatra but was born with a different de novo mutation. Hence, the descendants of Cleopatra can all be distinguished from the descendants of Cornelia, as they each have a different stamp in their mitochondrial passports. By looking at the particular series of passport stamps in one's mitochondrial DNA (the collection of mutations is known as a haplotype), we can form a complete picture of who descends from whom. Because people migrated around the world in different directions, their particular collection of passport stamps will vary by location, leaving us with a trail of breadcrumbs that allows us to track the ancestral record of our maternal ancestors. This trail extends all the way back to the so-called mitochondrial Eve, who lived in Africa around 170,000 years ago, from whom all humans descend.[16] Note that Eve was not the first human woman to ever live; she had many contemporaries, but her lineage outcompeted the lineages of other women.

A similar process occurs along our paternal ancestry. Whereas females inherit an X chromosome from each of their parents, males inherit an X chromosome from their mother, and a Y chromosome from their father. Because only men possess the Y chromosome, it too does not recombine and get shuffled with each generation; a man's Y chromosome is virtually identical to his father's, which is the same as his father's, and so on, back through time. Just as in mitochondria, there are mutations in the Y chromosome that get passed down through one's descendants and can serve as a series of passport stamps that identifies the ancestry of one's paternal line. Similarly, all men can trace their Y chromosomes back to a single individual—a Y-chromosomal Adam—who lived in Africa approximately 60,000 years ago.[17] (Note that this so-called Adam and Eve pair lived more than 100,000 years apart from each other).[18] Mitochondrial DNA and Y chromosomes can thus identify two branches of our genealogical tree that extend far back into time.

The third kind of ancestry information embedded in our DNA is quite different. This information is tied up in the other 22 pairs of our chromosomes (these chromosomes, which do not determine your sex, are called autosomes). Unlike Y chromosomes and mitochondrial DNA, the DNA in our autosomes gets reshuffled with each generation, so we cannot trace back an unbroken line of passport stamps. However, we can still learn a lot about our ancestry, despite this reshuffling, through a process known as genetic drift. Genetic drift occurs when random factors influence which people will reproduce, which accordingly influences which alleles will become more common in subsequent generations. The effects of genetic drift are most noticeable when people pass through a genetic bottleneck, which happens when only a small number of individuals come to populate a new region. An extreme example of this effect occurred when Fletcher Christian led a mutiny on *The Bounty* in 1790 in the

South Pacific and overthrew Captain Bligh. Christian and eight other mutineers, together with 19 male and female Tahitians who were with them at the time, sailed to the uninhabited Pitcairn Island, where they settled. Almost 70 years later, after the population of Pitcairn Island had outgrown its resources, a couple hundred descendants from this original founding population were resettled onto Norfolk Island near Australia. A recent analysis of Norfolk Islanders, almost all of whom can trace their ancestry back to the original Pitcairn settlers, finds that the population has unusually high genetic risk factors for cardiovascular disease, likely due to genetic drift.[19] That is, these deleterious genes were more common by chance among the founding population at Pitcairn compared with the rest of the world's population, and their offspring inherited these same genes. Because people differ from each other in how many offspring they have, genetic drift is always occurring, and it is a key reason why the frequencies of particular alleles vary around the world.

Because some alleles are more common in some parts of the world than in others, it is possible to identify the geographic origins of your ancestors by looking at the DNA in your autosomes. For example, the 5-HTTLPR region of the SLC6A4 gene (the so-called "depression gene" that we discussed in Chapter 4) commonly comes in two different versions, a short and a long allele. But the proportion of people who have each of these two alleles varies a great deal across countries. For example, the short allele is possessed by 80 percent of Japanese, 59 percent of Indians, 45 percent of Americans, and 28 percent of South Africans.[20] So if we look at a genome that possesses the short allele of the 5-HTTLPR region, and know nothing else about the owner of that genome, we would predict that the owner of the genome is more likely to be Japanese than South African. If we only look at this one allele, there isn't that much precision in our estimate of the person's ancestry—indeed, even though the short allele is far less common in South Africa than it is in Japan, there are still several million South Africans who possess it. However, if we look at a snapshot of thousands of alleles across the whole genome, we can make far more precise estimates of the likelihood of the person's geographic ancestry. For example, one study used a gene chip that assessed a half million different SNPs of a few thousand Europeans, and then came up with an estimate of their geographic ancestry based on the particular pattern of SNPs that they each possessed. This analysis was remarkably precise: about 50 percent of the people in the sample reported that they were born in a location that was within 310 kilometers of the location predicted by the pattern of their SNPs.[21] Other analyses can be even more precise: one study of people living in Sardinia was able to place half of them to within *15 kilometers* of their current home based on their distinctive genetic patterns.[22] Moreover, by studying the pattern of SNPs in your autosomes, it is possible to estimate the proportion of your DNA that comes from different regions of the globe. If one of your grandparents was of Nigerian ancestry, and another was from Ukraine, your own DNA would preserve signatures from each of these regions, and would indicate that upward of 25 percent of your DNA was Nigerian, and another 25 percent was Ukrainian, depending on how far back your grandparents' ancestry goes in those regions.[23] In contrast to mitochondrial and Y-chromosomal DNA, which only tell you about the ancestry of two branches of your vast family tree, autosomal DNA can estimate where many of your past ancestors have come from.

With all of this family history recorded in your DNA, would you like to get yourself genotyped to learn about your own ancestry? What you find out might provide you with a deep sense of connection to your past, or it might just completely upend your whole sense of who you are. Consider the genotyping experience of Wayne Joseph, a principal of a large suburban high school in Southern California. Joseph has been a pillar of the African American community there, and a staunch advocate for African American rights. He once published an article in *Newsweek* arguing that Black History Month only served to marginalize the African American community.

Like many African Americans, whose genealogical past is a mystery to them because of the slave trade, Joseph was curious about where his African ancestors came from.[24] After watching an episode of *60 Minutes* that discussed genetic ancestry testing, he decided to have his own DNA tested. He scraped his cheek with a swab and sent a sample of his cheek cells to DNA Print Genomics. When his results were sent back to him, Joseph learned that he was 57 percent Indo-European, 39 percent Native American, 4 percent East Asian, and *ZERO* percent African. After a lifetime of being African American, Joseph learned that he was disqualified by his genes.

Joseph experienced what sociologists call "genealogical disorientation"[25] and found his results completely discombobulating. "I kiddingly say, if I was 21 instead of 50, I'd be in therapy . . . it does rock your whole world."[26] After getting his results, he felt compelled to ask his mother, Betty, if he was adopted. "He is not adopted," she said. "Mother doesn't forget when she has a baby. And I had three babies. And he was one of them." Of course, ancestry testing is not infallible, and Joseph's friends have urged him to get retested. But as he says, "My response to them is, 'OK, let's say I get retested, I come back 9 percent African, 10 percent African, so I'm back in the club now?'" Just as with Szegedi, Joseph's life narrative was entirely upended by some unexpected knowledge about his genes. You might think that your sense of identity should be based on your life history and set of experiences—that is, it should arise from what you have done and the relationships you have. But as these examples reveal, for many people, their "real" identity comes from the origin of their genetic essence.

I found the siren's call of my own genome too irresistible to ignore, and I had my ancestry assessed by two companies: Genebase provided ancestry information about both my mitochondrial and Y-chromosomal DNA, and 23andMe provided information about mitochondrial, Y-chromosomal, and autosomal DNA. I was curious to see how well the genomic information would align with family lore. From what my family knows, all of our ancestors are culturally

German. My paternal grandmother immigrated to Alberta, Canada, from near the Alsace-Lorraine border between Germany and France, where she was raised by her German parents. My maternal grandmother's parents left a German community in Ukraine to move to the United States, en route to their final destination of Alberta, where my grandmother was raised.* And coincidentally, both of my grandfathers were born in Łódź, Poland, to German-speaking parents, and they too immigrated to Alberta, where everybody got together.

How well did the analyses of my DNA match our family story? Well, first, looking at my autosomal DNA, 23andMe identifies me as 100 percent European, and further breaks that down to 50 percent northern European, 31.3 percent eastern European, 10.7 percent southern European, 7.9 percent "broadly European," and < 0.1 percent "unassigned." At a finer-grained analysis, my three biggest sources of ancestry are 19.5 percent "French & German," 9.5 percent Scandinavian, and 7.3 percent Balkan. This all fits well with expectations, although I was surprised to see how small my percentage of German ancestry is, given that's the only cultural ancestry that I've ever heard about. Apparently, my distant ancestors are from all over Europe.

An analysis of my Y chromosome also fits with family lore. Genebase identifies me as part of the R1b haplogroup (a haplogroup is a group of people who share a haplotype), which the company describes as the dominant paternal family group of western Europe. Genebase further says that I must descend from a man, himself a descendant of Cro-Magnon people, who was born approximately 15,000 to 20,000 years ago in the Iberian Peninsula. I find this all vaguely self-validating, and in my mind's eye, I can see someone who looks like a hairier

---

* Interestingly, our family had a different version of my maternal grandmother's history, but I was contacted by a man through 23andMe, who the company identified as being a third or fourth cousin because of our shared genes, and he supplied me with the correct history of my maternal grandmother.

version of my grandfather walking through a dry forest wearing an animal skin. Although it was interesting to learn that one of the branches of my family tree extended back 15,000 years to southern Europe, it also became quite clear to me that information about this single ancestor really provided me with almost no information of any significance to me. I already figured I must be of European descent, and now I knew that one of the many branches of my family tree had been in Europe for a long time.

23andMe, which genotyped me in 2011, also analyzed my Y chromosome. The company also identified me as part of the R1b haplogroup, but looked at a broader series of markers than Genebase did, and could further specify that I was of the narrower haplogroup R1b1b2a1a1, which 23andMe says extends back 17,000 years to the most likely location of the fringes of the North Sea. Again, despite the precision, this information seemed to be of limited value—one of my distant ancestors was northern European, although I note that this was different from the southern European origins of my paternal ancestry, as Genebase had claimed.

23andMe also provided information about my mitochondrial DNA, and this was where things got more interesting. The results stated "Along your mother's line, you have ancestry in Eastern Asia in the past few hundred years" and showed a map with an arrow pointing to China. China! How was this possible? This didn't fit with any of our family stories. My great-great-great . . . grandmother came from China? What could this mean? My own essentialist biases kicked in and I caught myself wondering if this had anything to do with why I've always been so fascinated with Japan. I told my mother about 23andMe's findings, and I was surprised at how quickly she came to terms with it. Although she at first sounded taken aback, the very next thing she said was that she had always thought that her maternal grandmother had high cheekbones. Perhaps those were Chinese cheekbones! I also told one of my cousins about the test results of our shared mitochondria. She asked in disbelief "We're

Chinese?!" She later came up with a theory that seemed a lot more reasonable to her—our distant ancestor must have been a European who was living in China at the time. Maybe we were descended from Marco Polo's wife! With her version, the story of our European heritage could remain intact, with the additional twist that it involved a well-traveled ancestor.

However, a closer inspection suggests two key problems with my purported maternal line of Chinese ancestry. The first problem is that it's probably not true. My maternal haplogroup as identified by 23andMe is X2c1, which is a subgroup of haplogroup X. This is the most widespread haplogroup in the world, which more or less covers the entire Northern Hemisphere. Although it is possible that my maternal ancestry derives from China, the haplogroup doesn't have a definitive center of gravity, and it's far more statistically likely that it came from somewhere *outside* of China. It could really be from almost anywhere. But for some unspecified reason, 23andMe has opted to single out China as the likeliest origin of my maternal line.

In contrast, Genebase, which also identified my maternal ancestry as coming from haplogroup X, described my maternal ancestor from this broadly distributed haplogroup as likely coming from somewhere in Europe, the Near East, North Africa, the Druze in Israel, and throughout many Native American tribes. When I first got my Genebase results, all I could tell people was that my maternal ancestor came from somewhere in the Northern Hemisphere. Not surprisingly, most people laughed at the vacuity of this, and showed no interest in getting tested themselves. In contrast, when I tell people that 23andMe says my maternal ancestry derives from China, everyone seems fascinated, and many say that they want to get themselves tested too. 23andMe again shows its ability to deliver an exciting and engaging product, even if the information is sometimes embellished far beyond what is scientifically justifiable. This feels similar to the way that 23andMe presented my medical risk information—the company provided exact risk estimates past the decimal point for each common disease that it covered, even though there is no scientifically warranted basis for offering that kind of precision.

The second problem with noting that I have ancestry in China is that *of course* this must be true. You certainly do as well. The reason for this is simply the exponential way that the number of our ancestors increases as we proceed backwards in time. You have two biological parents, four grandparents, eight great-grandparents, and so on; the number keeps doubling every generation back through time. So if we assume that the average human generation is 25 years, then there have been approximately 32 generations since the Magna Carta was signed in 1215. This means that at the time of the signing of the Magna Carta, you would have had $2^{32}$, or approximately four billion, ancestors alive. An obvious problem with this calculation is that there were only about four hundred million people living on Earth back in 1215.[27] This means that many of the branches of your genealogical tree are crossed—some of your ancestors have interbred with your other ancestors.

Interbreeding ancestors aren't very common when we just look within the past couple of generations—the percent of the world's population who are the product of unions between cousins must be rather small (although in many smaller cultures,[28] such as the Samaritans, such marriages are common). But when we continue back to second cousins, third cousins, and so on, interbreeding becomes very common throughout the world. In fact, it quickly becomes unavoidable, and is something that geneticists refer to as pedigree collapse. The world's genealogical forest is not made up of neat rows of distinct individual family trees—it's a single giant mass of tangled brush. The extent of this tangling becomes evident when we look at those genealogical trees that have been doggedly scrutinized, such as those for various celebrities. For example, President Barack Obama is purportedly a distant relative to President George W. Bush (an eleventh cousin), Vice President Dick Cheney, Sarah Palin, Rush Limbaugh, Warren Buffet, and Brad Pitt.[29] It's not that

Obama uniquely descends from some kind of celebrity gene pool—it's just that if you look far enough back in time, you'll discover how interrelated we all are.

The further we go back into the past, the larger the percentage of people alive at that time were your direct ancestors. And, likewise, the deeper we go back into time, the greater the likelihood that you and I share a common ancestor. At some point in the past, one of the branches of my exponentially growing family tree intersects with one of the branches of your equally vast family tree. For example, just looking within Europe, one analysis of the genomes of a large sample of Europeans found that people who lived in opposite corners of Europe were likely to share *millions* of common ancestors over the past 1,000 years.[30] Given how interrelated humans are, at some point in the past, there must have been a person that all of humanity can count as their ancestor. We saw with Y-chromosomal Adam that everyone's paternal line of descent leads to the same man who lived in Africa approximately 60,000 years ago. But note that this is just a single branch of your vastly convoluted family tree—the branch leading to your father's father's father's father. . . . If we looked across *all* of the branches of our family trees we must share a common ancestor that is far more recent than this. How far do we have to go back in time until we reach an individual who is an ancestor of all of us through any branch of our family trees? One statistical analysis, which took into account people's continent of origin, historical population movements, and the frequency with which people tend to reproduce with someone from outside of their town of birth, estimates that the most recent ancestor of all of humanity lived around 55 CE.[31] Assuming 25-year generations, this is about 79 generations ago, which would mean that the most distantly related person to you on the planet is something like your 78th cousin. Moreover, taking things even further, we can identify the point in time at which all of the people living today share the exact same set of ancestors who were alive back then. This date is estimated to be 2158 BCE. This means that everyone in China, or anywhere else for that matter, who was alive in 2158 BCE, and whose lineage has been fortunate enough to survive to the present, was one of your ancestors and one of mine as well. This also means that the man who was walking in the Iberian forest 15,000 to 20,000 years ago that Genebase identified as the origin of my Y chromosome is your direct ancestor as well, although probably not through the same particular line of descent as me. Based on these kinds of statistical analyses, the science writer Steve Olson conjectures that all humans who are alive today are likely direct descendants of Confucius, Nefertiti, and Julius Caesar, given that each of them left surviving offspring more than 2,000 years ago.[32] Olson further speculates that if Jesus had any children, which some argue that he did,[33] we may all be descendants of him too.[34] So by this accounting, I must be Chinese after all, as well as Egyptian and Roman, and maybe I even have a little bit of Jesus in me—just like you. We are each made up of tiny bits of DNA that originated from across the whole world.

That the branches of our family trees get increasingly entangled as we go further back in time reveals an inherent paradox in our fascination with genealogies. We look to genealogy to see how our own distinctive life story is built upon the unique collection of essences that have been passed down to us over the ages. Genealogy is so attractive because it provides us with a narrative that extends beyond our own lifetimes and connects us to people in the past. These connections can change the ways that we feel about ourselves. The psychotherapist Anne Ancelin Schutzenberger even argues that our ancestors' collective memories somehow get passed down to us together with their genes, leaving us to unconsciously struggle with their unfinished business in our own lives.[35] This idea of a kind of genealogical reincarnation remains a relatively popular and intuitively satisfying notion (it's even the underlying premise of the popular video game *Assassin's Creed*), despite there being no science to back it up.

The challenge with trying to understand ourselves through genealogy is that the further we go back in time, the more diluted are our connections with any individual people who were alive back then, and the more our set of ancestors becomes similar to everyone else's set of ancestors. Despite how little a distant ancestor may have actually contributed to our genome, however, that connection can still have a pronounced psychological impact upon us. For example, I imagine that considering that your great-great-great . . . grandfather could be Jesus might make you pause and think a bit. The feeling that your own life story connects to other individuals is one of the key psychological rewards of genealogy.

The impact that genetic ancestry testing has on people has been investigated by my colleague, the University of British Columbia sociologist Wendy Roth. She has been exploring how people respond to the information about their own ancestry from direct-to-consumer genomics companies. Roth contacted more than a hundred people who had taken a genetic ancestry test to interview them about their experiences. The information that they received had a striking impact on many people. In particular, people's genetic test results frequently caused them to identify their race or ethnicity differently from before—indeed, a full 40 percent of Roth's participants reported a change in their identification.[36] Many people found their genetic ancestry results to have a positive change on their identity. One woman who had previously identified as having Scottish and Native American ancestors seemed quite thrilled to have her Native American ancestry confirmed. "Before we just had family stories of [an] Indian princess, now I know I have Native American genes and am proud of it. . . . [I'm] planning on identifying myself as Native American on 2010 Census this month." Likewise, an American man of European, Chinese, and Mexican ancestry, and who teaches at a Black college, was quite happy to learn that he also had African ancestry. He could use it as a means to build connections with his students: "[they] sometimes pay more attention to me when I mention my ancestry when discussing racial/ethnic topics." Another

White woman who learned of her African ancestry ended up feeling a new sense of connection with African Americans: "I tended to view African Americans as a different group of people, unrelated to me. I now know I am related to many African Americans and they are related to me."

But receiving results that conflicted with one's expectations also created a lot of consternation in some, much like the experiences of Wayne Joseph. One man, who had always identified as Anglo-Saxon, was surprised to learn that he had substantial amounts of German ancestry, and responded, "I don't know who to support in the World Cup now." A White British man seemed to be struggling with his results and said he was still "coming to terms with being African." And the reaction of a White Californian man who was informed that his paternal ancestry came from Africa responded by saying "As that doesn't seem to make any sense at all, I'm having another test done by a different company." Perhaps this is part of the business model of the direct-to-consumer genomics industry—hoping that people will keep retaking the tests until they get the answer that they're looking for. For his part, Wayne Joseph is no longer a fan of genetic ancestry testing, and he warns people to stay away from these tests precisely because of what they might find out: "You don't want to know," he says. "It's like a genie coming out of a bottle. You can't put it back in."[37]

Discovering where their genes originate doesn't just impact how people think about themselves—approximately half of the participants in Roth's study said that their genetic test results affected their activities and friendships. For example, a Kansas woman who had previously identified herself as of White/Hispanic descent set out to make a lot of new friends in various Native tribes upon learning of her Native American roots; "[my] new friends who are enrolled in various tribes . . . look at me as an equal." A California man who had believed that his ancestry came from Wales and England felt validated when he learned that his genes did indeed derive from

Wales: "It makes me feel much closer to Wales and I read their news and have listened to the Welsh language and been curious about learning it."

We can also see this impact on one of the most famous people to have their genetic ancestry tested: Oprah Winfrey. In the summer of 2005, Oprah went to South Africa, where she announced: "I always wondered what it would be like if it turned out I am a South African, [because] I feel so at home here. . . . Do you know that I actually am one? I went in search of my roots and had my DNA tested, and I am a Zulu."[38] So moved was she by her newfound sense of Zulu identity that Oprah devoted her efforts to building a leadership academy for girls in South Africa. Oprah later agreed to have her ancestry tested again as part of a PBS series, *African American Lives*. This time, though, she was informed that her DNA had three exact matches: the Kpelle people of present-day Liberia, the Bamileke people in Cameroon, and Nikoya people from Zambia[39]—this second test revealed no Zulu connections at all. So the question of where Oprah's ancestors actually come from is still quite up in the air.

Unfortunately, Oprah's experiences with clashing ancestry test results aren't that unusual. Similar to the conflicting results that I received from Genebase and 23andMe, this suggests that there is some overpromising in the precision of genomic ancestry test results. Several researchers have issued harsh critiques of the genetic ancestry testing industry because there are many limitations to the enterprise that are not made clear to customers.[40] First, the reference database to which your genes are compared is composed solely of genetic data from contemporary groups, but you are given predictions about which *ancestral* groups your genes come from. There are many ways that the ancestral populations from particular regions differed from those living today, meaning that our understanding of the genetic composition of ancestral populations is built on a significant leap of faith. Moreover, the existing reference database has far too few samples to allow for precise predictions, and there are many populations around the world that have not yet been sampled, meaning that the database remains somewhat nonrepresentative. In addition, the analytic techniques that each company uses to make predictions are not made transparent to consumers, and there is really is no governmental oversight to ensure that companies are providing accurate information.

The most glaring problem, though, is that estimates of your genetic ancestors are always probabilistic, and are not anywhere near as definitive as the optimistic claims from direct-to-consumer genomics companies suggest. 23andMe bluntly tells me that "Along your mother's line, you have ancestry in Eastern Asia in the past few hundred years." This bald claim of fact makes no effort to convey all of the hedging and uncertainty that this prediction is built upon. These companies provide you with a best guess of where your ancestry comes from—they can't identify any conclusive places of origin—and often different companies come up with quite different guesses. Given how much people are affected by any information of where their essence comes from, they typically will latch on to any story of ancestry that they are provided. Yet it's quite likely that if people have their ancestry tested by a different company, they may encounter an entirely different story.

Perhaps we shouldn't be overly concerned if the precision of genetic ancestry testing is exaggerated because there isn't all that much at stake in being curious about your genealogical roots—unless you're planning on building schools in your ancestral homeland, like Oprah. The kinds of genetic markers that are used to determine your ancestry are, for the most part, of no biological significance. At a biological level, it really doesn't matter if my mitochondria come from China, or if Wayne Joseph's DNA doesn't come from Africa. The significance is all at the psychological level, so maybe—as long as you just get some kind of concrete answer to your estimated geographic origins—your genetic ancestry results don't need to be accurate. However, the costs of the overpromising of tracing DNA ancestry are more pronounced when they are used to adjudicate legal matters. For example, genetic ancestry tests have been turned to for evaluating whether one can claim tribal citizenship within certain Native American tribes,[41] or for making claims to qualify for college minority scholarships.[42] The United Kingdom Border Agency launched a controversial pilot project in 2009 when it sought to use genetic testing in the hope of discriminating between those Somalian refugees who were legitimately seeking asylum and potential opportunists from neighboring African countries. The program was abandoned shortly thereafter when it came under attack for not being able to offer the precision necessary to make any valid distinctions in the asylum-seekers' origins.[43] As more and more people become familiar with DNA ancestry tests, it's likely that people will turn to these tests to address other kinds of legal issues. It's essential, therefore, to recognize just how imprecise these estimates of genetic ancestry really are.

## Essences and Race

The psychological impact of learning about how genes are dispersed around the world extends beyond your own sense of identity. It's also an issue for how we think about that most contentious of topics: race. As we noted in Chapter 3, essences are seen to carve nature at its joints. Shared essences are what bind species together, and differing essences are what distinguish species from each other. And we extend this same way of thinking about essences to our own species.[44] For example, if you reflect on the shared genetic heritage of people living in Japan, then you may be more likely to feel that, in some ways, all Japanese are alike—that there's something inherently Japanese in everyone who shares that same genetic heritage. The flip side of this also seems to be true, and this is where some problems can arise. If you reflect upon how others possess an essence different from your own, they will likely seem fundamentally different to you. The essence that makes you who you are is not shared by them—there is now a line drawn in the sand that distinguishes a genetic "us" from a "them." So, if you're not Japanese yourself, and if you reflect on how Japanese all seem to share a common genetic foundation, then you'll probably feel more different from them. Perhaps, you may also come to feel more prejudice toward them as well. Racist thinking is built upon this idea of people with discrete and nonoverlapping essences.

With this in mind, there are two very different scientifically justified stories that we can tell about the distribution of the human genome, and each of these has far-reaching consequences for the ways that we think about other people. Story One speaks to how genes can divide the world. It's based on the notion that people who think they possess different genes from others will feel more prejudice toward them. Story Two, on the other hand, is more optimistic, and it speaks to how genes can unite us all. Reflecting on the genes that we have in common with people from other parts of the world can lead to a reduction in racism. Let's explore each of these stories and their psychological aftermath.

## Story One: Genes Can Divide Us

What happens when people reflect upon the fact that genes are distributed differently around the world? The social psychologist Johannes Keller explored this question by asking German students to read one of two different essays.[45] The first essay described the science about the distribution of genes around the world, very much like I described it earlier in this chapter. That is, the students read about how certain genes are distributed unequally around the world and that you can use new genotyping technologies to estimate the ancestry of people from around the world. A second group of students read an essay that had nothing to do with genetics—it was a control essay that focused on entirely unrelated matters. Following the essay that they read, the students were asked some questions about a controversial political topic regarding the expansion of the European Union. They were asked to indicate their feelings toward citizens of several western European countries (such as France, Germany, and Italy) and to those of several eastern European countries (such as Bulgaria, Poland, and Turkey). Keller then compared how warm people's feelings were between the different countries depending on whether they had just read about the geographic spread of genes or the control essay. What did he find? Well, it turned out that those students who were contemplating how genes were different around the world had considerably more negative attitudes toward eastern Europeans than did those who read the control essay. The essay seemed to underscore that if genes are different around the world, then eastern Europeans really *are* different from western Europeans at a fundamental level. And when that was brought to the forefront of the German students' minds, they showed more prejudice toward them.

These findings fit in well with the way that psychologists have traditionally understood prejudice. We tend to understand other groups by imagining that there is some kind of mysterious essence that makes them as they are. Here is how one of the first psychologists to study prejudice, Gordon Allport, described in 1954 how essences were key to understanding why people feel prejudice: "There is an inherent 'Jewishness' in every Jew. The 'soul of the Oriental,' 'Negro blood,' Hitler's 'Aryanism,' 'the peculiar genius of America,' 'the logical Frenchman,' 'the passionate Latin'—all represent a belief in essence. A mysterious mana (for good or ill) resides in a group, all of its members partaking thereof."[46] Genes are a perfect stand-in for this shared mana or essence, and when we reflect on them, we become more prejudiced.

For example, I and my students Benjamin Cheung and Cermet Ream gave a large sample of American adults a number of different questionnaires. Some of these tapped into how much people believed that genes determined life outcomes. For example, people were asked how much they agreed with statements such as "An individual's particular behavior is not changeable if it has a genetic basis."[47] And they also completed some other questionnaires that assessed various kinds of racist thinking. For example, one survey assessed whether they believed that some kinds of people were simply less deserving than other kinds of people. It included items such as "Some people are just inferior to others."[48] Another survey tapped into their feelings of racism against Blacks in the United States. It included items such as "Over the past few years Blacks have gotten more economically than they deserve."[49] And yet another survey tapped into how much people felt that those who challenged the status quo should be punished. It included such items as "What our country really needs instead of more 'civil rights' is a good stiff dose of law and order."[50] What we found was that for all of these different surveys of prejudice and intolerance for outgroups, the more people thought that genes were like essences, the more negatively they viewed outgroups and those who stood to challenge the status quo (at least, for those who were not African Americans themselves).[51] Thinking about genes as underlying essences and having racist thoughts seem to go hand in hand.[52]

## Genes as the Ultimate Causes of Racial Differences

Story One thus leaves us with the impression that different genes around the world have created discrete races that will forever be in opposition because we're all ultimately made up of different stuff. But Story One has an even more nefarious consequence. Our tendency to see genetic groups as homogenous and discrete connects with our penchant to think of genes as ultimate causes. This means that we find it easy to assume that any group difference that exists must be the result of the different genes possessed by people who belong to that group.

There are many anecdotal examples of this bias. The co-discoverer of DNA James Watson has repeatedly gotten himself into trouble for suggesting that genetic differences underlie cultural differences. For example, he has argued that skin color was associated with enhanced sex drive: "That's why you have Latin lovers. . . . You've never heard of an English lover—only an English patient." Watson was ultimately forced to leave his job as chancellor of Cold Spring Harbor Laboratory for suggesting that African genes left those in that continent less intelligent than elsewhere.[53] Likewise, Nike encountered some backlash after it released the Nike Air Native—a shoe supposedly designed, in Nike's words, to fit "the distinct foot shape of American Indians."[54] This campaign was misleading in its assumption that Native Americans all possessed the same foot shape, which was different from that shared by the rest of the population.

When I lived in the small town of Obama in Nagasaki prefecture, I attracted a lot of attention as the first Westerner to have ever lived in the town, and many people I met sought to explain my own cultural idiosyncrasies in terms of my underlying biology. For example, a frequent observation that people would make to me was to note that Japanese people ate rice, whereas foreigners ate bread—a conversation topic that always seemed to come up when I was eating rice. And some would go further to offer an explanation for why this cultural difference existed: it's because the intestines of foreigners were too short to properly digest rice. This idea that foreigners have different intestines was quite widely shared in Japan; in 1988, the Japanese Ministry of International Trade and Industry made the official case that American beef should be kept out of Japan because it was ill-suited for the makeup of Japanese intestines.[55] Another person once explained to me that the reason why pachinko (a cross between a slot machine and a pinball machine) was popular in Japan but not in the West was because foreign thumbs were the wrong shape to turn the dial properly. My freakish foreign biological makeup often seemed to be on people's minds when they interacted with me: someone once even expressed surprise that my nose was congested when I had a cold, given just how huge it was.

A textbook example of genes being turned to as the ultimate explanation of cultural differences was made in *A Troublesome Inheritance*, the controversial 2014 book by the former *New York Times* science reporter Nicholas Wade. Wade tries to make the case that race is very much a biological phenomenon and the reason that people do things differently around the globe can be understood as a direct function of their distinctive genes. So Wade opined that the Middle East remains ungovernable because people there possess tribal genes, Chinese rarely make innovations because they are saddled with conformity genes, the Industrial Revolution began in England because the English had productivity genes, and Africa's economic problems are rooted in the fact that their genes never evolved to live in complex urban environments. Wade does at least acknowledge up front that these conclusions are all speculative, which is fitting as he doesn't offer any kind of genetic evidence that might support them. Geneticists certainly don't agree with his interpretations of genetics research: a letter denouncing Wade's review of the genetics literature was signed by more than 140 population geneticists—many of whose research Wade cited in building his arguments.[56] The notion that we need to turn to genes to account for cultural differences is especially unconvincing, given that there have been several fields of inquiry that have documented how cultural, economic, geographic, and political factors contribute to population differences.[57]

Perhaps it's plausible to speculate that genes that contribute to various kinds of valued social characteristics might be distributed unevenly around the world—a point that we'll return to in the next chapter. But one challenge with using genes to explain group differences is that it all depends on the point in time that you're looking back from. As Jonathan Haidt and Steven Pinker point out, during the 1930s, it would have seemed quite obvious to look for war-mongering genes in Germans and peace-seeking genes among

Ashkenazi Jews to account for the striking population differences that were evident at the time. But political events in the 21st century would instead have us looking for the precise *opposite* kinds of genes in German and Israeli populations.[58] Likewise, speculating that Chinese lack genes for innovation might be consistent with their smaller number of Nobel Prizes, but in the 15th century, the Chinese were far ahead of the West in scientific progress.[59] But given our essentialist leanings, for many people the only satisfying explanation of group differences is to posit an imagined essence that gives rise to these differences.

There are even examples of the medical community blithely assuming genetic differences underlying any health disparities across the races.[60] For example, consider the well-studied case of hypertension. By the 1960s, it had become clear that African Americans had far higher rates of hypertension than did White Americans; for example, 38.6 percent of African American women have high blood pressure compared with only 22.6 percent of White American women.[61] What might be the cause of this racial difference? You probably won't be surprised that the most popular hypothesis to emerge was that this difference must be genetic in origin. Several researchers posited the existence of Black hypertension genes, and this led to a number of expeditions in search of these problematic genes. However, this endeavor soon had to come to terms with the awkward realization that hypertension rates were much higher among African Americans than they were among people in West Africa, where those Black hypertension genes must have originated.[62] This subsequently led to the more nuanced hypothesis that there must have been rampant natural selection operating during the slave trade that led to different genes becoming common among African Americans compared with West Africans. The so-called "slavery hypothesis" posited that the only Africans who were able to survive the brutal conditions of the Middle Passage as they were shipped from Africa to the New World were those who possessed salt-retention genes.[63] Although those genes may have been adaptive on the Middle Passage, in the salt-rich life of modern America, those same genes are seen as the cause of the higher hypertension rates among African Americans. This slavery hypothesis was widely covered in the popular press, but however intuitively appealing this theory may be, it has come under harsh criticism for want of supportive evidence.[64] Yet despite this stark lack of empirical evidence, the theory remains widely discussed and embraced to this day. As epidemiologists Jay Kaufman and Susan Hall note: "The intellectual resilience of the Slavery Hypothesis may be attributable to . . . the beguiling allure of a simplistic genetic determinism."[65] Our search for essences as the ultimate causes means that the most intuitive explanation for why more Black Americans suffer from higher blood pressure than White Americans is that their genes are different. Essence-based explanations can be so attractive that they can lead us to miss the far better empirically supported cause of this racial difference—namely, that the higher blood pressure among African-Americans has something to do with the challenges and stressful life experiences of being Black in America.[66]

## Story Two: Genes Can Unite Us

I have to admit, Story One is quite a downer. When we focus on how genes are distributed differently around the world, it seems to stoke our deepest prejudices. Given that there really are distinctive patterns of genes that are dispersed across the globe, it's discouraging to see how people respond to this truth about human genetic diversity. But there's a very different story that we can tell about the diversity of human genes: Story Two focuses on how our genes serve to unite all of humanity together. There are two key parts to this story. First, Story Two highlights the social reasons for why the races appear different; and second, it reveals that there simply isn't that much human genetic variation to begin with. Let's discuss each of these in turn.

## Biological versus Social Accounts
## of Racial Differences

Research on genetic ancestry testing has confirmed that there are genetic markers that reliably distinguish people from different parts of the world. And often we can even guess with some reliability someone's ancestry just by looking at their physical features. Surely, no one has ever mistaken Donald Trump for an African, for example. Likewise, you might have noticed how people from different regions of the globe are overrepresented in some occupations compared to others; for example, there is a much larger percentage of people with African ancestry playing professional basketball than there are playing professional hockey. Observations like these can make it seem that the whole concept of race is really a statement about genetic differences across continents. Is race simply the product of our biology?

Although there are clear patterns of genetic variation among humans across the globe, by itself this doesn't necessarily tell us anything other than that humans have traveled the world over and brought their continually mutating genes with them.[67] To say that race has a meaningful biological basis requires specific kinds of evidence. Biologists apply specific criteria when they are determining whether different "races" exist in other species (although the term that is usually used for nonhuman races is "subspecies"), and we can apply those same criteria to the human case.[68] For example, biologists agree that a species can be said to have different races when there is *evidence of a sharp genetic boundary* distinguishing one population from the other.[69] But for humans, there are no clear-cut boundaries; rather, for any given gene, there are gradations of differences across populations.[70] Gradations of genetic differences are what you'd expect if people were reproducing largely with those in their own population and occasionally with people from nearby populations, as humans are wont to do.

Relatedly, a second test that biologists apply to determine the existence of races is to assess the overall amount of genetic differences between populations. The threshold is set that *at least 25 percent of the genetic variance between individuals must be between populations* for them to be identified as belonging to different races.[71] That is, if there are, say, 1,000,000 genetic variants that differ between Donald Trump and Oprah Winfrey, we would need to see at least 250,000 of those variants differing between people of European descent and people of African descent more generally in order to classify them as belonging to different biological races. For humans, however, only 4.3 percent of genetic variation between individuals lies between the continental "races" of humans[72]—the other 95.7 percent is due to reasons other than their race. That 4.3 percent is not meaningless— much of that genetic variation leads to the differences in appearances (such as skin color, hair, facial features) that allow us to say with confidence that Donald Trump and Oprah Winfrey look like they belong to different races. But this relatively small number tells us that the vast majority of genetic differences among individuals has nothing to do with race.

A third test for being able to identify a race is that races need a *distinctive evolutionary lineage*, meaning that almost all interbreeding is done within populations rather than between them. Because throughout history, humans have always had much interbreeding across populations, especially between neighboring groups, they roundly fail this test.[73]

Finally, races can be said to exist if they have *uniquely evolved adaptive traits* that are different across populations. An example of such an evolved adaptive trait in humans is skin color; as we discussed earlier, skin color evolved to adapt to the level of ultraviolet radiation at different latitudes. However, the adaptive traits of different populations are not unique to just those populations. For example, sub-Saharan Africans and Melanesians in New Guinea both have similarly dark skin color despite being among two of the most genetically distinct populations on the planet.[74] Typically, we identify different races as existing on the basis of visible markers, such as skin color and facial features. But these visible markers are just the tip of the genetic iceberg. There is far more adaptive genetic variation between populations beneath the skin that we cannot see—for example, genes underlying a resistance to malaria, or an ability to digest lactose as an adult. But these genetic differences do not cluster in any sensible patterns that resemble what we think of as the major continental races.[75] Moreover, skin color and facial features that we use to distinguish races do not correlate well with other aspects of our biology[76]—for example, West Africans with genes that lead to curlier hair are not any more likely than other West Africans to have genetic resistance to malaria.

Again, it's important to remember that the vast majority of human genetic variation is due to genetic drift,[77] and most of the genetic variants dispersed through genetic drift have no discernible impact on our phenotypes. That is, although genetic ancestry testing can allow us to predict our ancestry by patterns of genetic variants around the world, this does not mean that those same genes correspond to any meaningful difference in our biological makeup.

In sum, the genetic variation among humans fails each biological test of being a race. This is why social scientists and geneticists are largely united (contra Nicholas Wade)[78] in the view that there is not a sound biological basis to the concept of race; rather, race is typically understood as a social construction—a product of what we learn growing up.[79] Societies decide who to classify as what, and what they decide varies from one place to another. In contrast to a biological account of race, this social account focuses our attention on the ways that our *experiences*, not genes, differ across the races. And unlike genes, experiences don't lead us to think about essences.

How can we see the social basis of race? Consider what sometimes happens to individuals who move to different countries. The South African comedian and host of *The Daily Show*, Trevor Noah, notes how the United States and South Africa disagree about who he is. Because he has a Swiss father and a Black South African Xhosa mother, South Africa sees him as colored. In contrast, in the United States he is recognized as Black. Similarly, many people who would be classified as White in Brazil would be classified as Black in the United States, reflecting the differing degrees of admixture of ancestries between the two countries.[80] Wayne Joseph's identification as African American in the absence of any African genes is also a strong example of how the notion of race is socially constructed above and beyond any actual genetic basis.

Another way we can see the social foundation of race is to consider President Obama, born to an African father and a White American mother. Why is he thought of as the first Black president and not the 44th White one? What makes the ancestry of his father count for more in his racial identity than that of his mother? This reflects the principle of hypodescent—the practice of assigning to children of mixed ancestry the ethnicity of their subordinate group. Hypodescent is a racial version of the negativity dominance principle that we considered in the last chapter. Bad is stronger than good, and in a racially hierarchical society—sadly, we still do live in societies where some racial groups have more status and power than others—the contributions of what some see as your lower-ranked ancestry will weigh more heavily in how people view you than your higher-ranked ancestry. The principle is evident in the writings of the eugenicist Madison Grant, who wrote in 1916: "The cross between a white man and an Indian is an Indian;[81] the cross between a white man and a Negro is a Negro; the cross between a white man and a Hindu is a Hindu; and the cross between any of the three European races and a Jew is a Jew."[82] The most extreme form of hypodescent in the United States emerged in the "one-drop rule." This rule, which first became codified into law in Tennessee in 1910, and was then adopted by several other states, specified that any individual with at least "one drop" of Black blood should be considered Black. And, as in the case of President Obama, people still show evidence for hypodescent today.[83] Hypodescent reveals the stark power that people grant to inherited essences; just the slightest bit of an essence from a lower-ranked group may be enough to categorize someone as belonging to that group. Hypodescent, and more generally the social basis of race, shows that it is society's views about certain ethnic groups that largely determine life outcomes, rather than the direct unfolding of any genes from those particular ethnic groups.

Does it really matter which theory of race, a biological or a social one, that someone subscribes to? Consider what happens when people are exposed to one or the other theory. In one study, White and Asian college students from California read one of two different fabricated newspaper articles. One of the articles made the case that race was the product of our biology. The article quoted a geneticist saying that "In the end, we obtain our genetic material from our parents, so we generally inherent their race along with everything else." The other group of students read a newspaper article that instead argued that race was the product of social norms. It argued that "the practice of classifying people into racial groups based on certain patterns of physical appearance is entirely cultural in origin."[84] The participants then watched a video of a Black student who described how he had been inexplicably fired from his job: "I've been working there for six months. I thought everything was going well. . . . I don't really know what happened." The participants were asked questions about how much they would like to become friends with this student. The results of the study were quite stark: those who had read the essay about the social basis of race were far more interested in becoming friends with the Black student than those who had read about the biological basis of race. If your race is seen to stem from your biology, then maybe the fact that this Black student had gotten fired revealed some inherent racial inadequacy in him that was causing problems. In contrast, if race is a social product, then the Black student's situation elicits far more sympathy.

Other studies have yielded conceptually similar findings.[85] For example, in another study, Asian American college students were also asked to read an essay that argued for either a biological or a social basis of race[86] and then were asked how much they identified with American culture. Those Asian Americans who had reflected on a social theory of race felt far more connected to American culture than those who reflected on how their race was grounded in their biology. It appears that those who thought of themselves as biologically Asian concluded their differences were immutable, and they would therefore never fit in well with American culture.

## How Genetically Different Are Humans?

The history underlying the prejudice-fanning flames of Story One originates tens of thousands of years ago, when some of the first modern humans left Africa and carried their continually mutating genes with them as they headed off to settle the distant corners of the world. However, we can tell an entirely different story that begins at this same historical juncture. Prior to leaving Africa, early humans must have faced some severe environmental challenges, as researchers estimate that the entire population of modern *Homo sapiens* was reduced to a few thousand individuals at this time.[87] It was the descendants of this relatively small group of people that ultimately populated every habitable corner of the globe.

This fact, together with our globe-trotting habits, our cultural inventions of ocean-faring ships, and our apparent willingness to reproduce with abandon wherever we went, has resulted in the collective human genome being rather distinctive in terms of just how massively similar it is from one person to another, regardless of an individual's ancestry. As discussed above, all humans today are likely related within about 79 generations of each other. There are noteworthy patterns of genetic markers around the world, but in absolute terms, there just isn't all that much genetic variability among humans. It's certainly much less than what is found in most other species.

Take our closest related nonhuman species: chimpanzees. Humans last shared a common ancestor with chimps about 6.5 million years ago.[88] Unlike humans, chimps didn't have the same penchant for globe-trotting; chimps never left Africa, and all chimps in the wild today are located within a fairly small belt that straddles parts of western Africa. Yet, despite this much narrower geographical range, the different populations of chimps have far more genetic variability between races when compared with humans. Remember that between two people from different continental races, only about 4.3 percent of their genetic variability lies between the races. In contrast, 30.1 percent of the genetic variability among individual chimpanzees lies between the different races[89]—that is, if you compared the DNA of a chimp who belonged to the race of "western chimps" with that of a chimp who belonged to the race of "central chimps," about 30 percent of their genetic differences would be shared with other chimps of their same race. This is because chimps didn't go through a recent genetic bottleneck the way humans did, and because chimps don't swim or have boats—they never get to hook up with the chimps that live on the other side of the river.[90]

Indeed, one of the most noteworthy surprises to come out of the sequencing of the human genome was that humans are so overwhelmingly genetically uniform. When President Bill Clinton announced the preliminary results of the Human Genome Project in 2000, he declared: "Today, we are learning the language in which God created life. . . . I believe one of the great truths to emerge from this triumphant expedition inside the human genome is that in genetic terms, all human beings, regardless of race, are more than 99.9 percent the same."[91] Clinton's take-home message for the sequencing of the human genome was effectively to argue that all humans share the same essence.

If our essentialist biases lead us to react to the idea of genetically different ancestries by bringing out our darkest prejudices, how will our focus on essences lead us to react to the notion of a largely uniform human genome? Clinton seemed to have an intuition: he viewed the remarkable DNA similarities around the world as having social significance for international relations. He once lectured to the Serbs and Croats in Kosovo about the inherent irrationality of their ethnic conflict, given their genetic similarities.[92] Was Clinton onto something? Does reflecting upon our genetic similarities make us more tolerant of people of different ethnic groups?

Some recent research supports Clinton's intuitions. In one study, Jewish American participants were assigned to read one of three different bogus news articles.[93] One of the articles reported on scientific research that highlighted how genetically similar Jews and Arabs were. It claimed that "Jews and Arabs are 'genetic brothers.'" A second article made the opposite point, and highlighted the genetic distinctions between Jews and Arabs. And a third article was on an unrelated topic. After the participants had read one of these articles, they were asked questions regarding their antipathy toward Arabs, indicating how violent, unfriendly, or unhelpful they found them. Then, the participants were asked how much they supported peacemaking efforts in the Middle East. Specifically, they were asked whether they agreed with the statement "In order to achieve its goals, Israel should pursue peaceful diplomacy with the Palestinians instead of using aggressive actions." The results indicated that those who read about the genetic similarities between Jews and Arabs were more in support of peacemaking efforts than those who had read the other essays. Likewise, those who had read the genetic similarity article reported less antipathy toward Arabs than those who read about genetic differences between Jews and Arabs. Focusing on our common genetic foundation might lead us on a path toward peace.

More generally, it seems that Story Two leads us away from relying on genes to make sense of our world. This became clear in a study that I and my students Anita Schmalor and Benjamin Cheung conducted about people's attitudes about ethnic differences.[94] We had a large sample of American adults read either an essay that was written to model Story One—it argued that you can estimate one's geographic ancestry through one's genes—or an essay about Story Two, which argued that the amount of genetic variation in humans is trivial compared to other species. And then the participants were asked questions about a long list of stereotypes, such as "Why do Africans tend to have a good sense of rhythm?" or "Why do Japanese have the longest average lifespans in the world?" What did we find? Well, those who read the essay about Story Two were more likely than those who read Story One to see these stereotypes as the result of people's experiences, as opposed to their genes. When we dwell on the common genes that humanity shares, we no longer expect that ethnic differences are the result of different underlying genetics.[95]

Story Two shows that overall, race is largely a social construction, and we share the vast majority of our genes with all our fellow humans. These facts are encouraging for the future of race relations, and I hope that learning about Story Two has a similar impact on the readers of this book. But as encouraging as this is, we still remain vulnerable to the allure of Story One. We can easily see cultural differences around the world—people from different places often look, act, and think differently from each other—and our essentialist biases make us prefer genetic explanations for those visible differences. This attraction to genetic explanations for our differences has led to one of the most destructive ideas the world has ever seen: eugenics.

# 7.

# Essences and the Seductive Allure of Eugenics

**IT'S HARD TO IMAGINE A MORE UNIVERSALLY** despised idea than that of eugenics. Literally meaning "good in birth," this is the philosophy which undergirded the Holocaust, and the Nazis' effort to rid the world of those who they saw as "undesirable." The logic behind eugenics is simple and is succinctly captured in the following statements: "the multiplication of the feeble-minded . . . [was] a very terrible danger to the race," or "Some day we will realize that the prime duty, the inescapable duty, of the good citizen of the right type is to leave his or her blood behind him in the world."

But, given how much eugenics is associated with the Nazis, it might come as a surprise to learn that these two statements came from Winston Churchill and Teddy Roosevelt, respectively. For despite how contemptuously most people today think about eugenics, in the first half of the 20th century, it was widely embraced throughout the developed world and by many leading thinkers and geneticists at the time. The allure of eugenics illustrates the power of essentialist thinking perhaps more pointedly than any other topic in this book. The extensive eugenics movement of the early 20th century rested on a search for the essences underlying two characteristics: intelligence and criminality. We'll explore how our essentialist biases are interwoven into the eugenics movement, first by considering the role of genes and essences in intelligence and criminality, and then by discussing how an attraction to essences guided the eugenics movement.

## Intelligence and Essences

So what is your IQ? Do you even know? A telling fact about the anxieties that intelligence testing entails is that it's quite possible that you were never told your IQ, even if it was formally assessed by a psychometrician (and don't trust any of the IQ tests on the Internet—they're generally not valid). If you were tested at school but were never told your results (as was the case for me), then this may very well be the only intellectual test that you've taken in which your score was kept hidden from you. It never appears on school transcripts, and it is almost certain that your IQ score was never made public to anyone outside your family and schools. In some ways, IQ scores are treated like state secrets. Why would there be this secrecy around the results of IQ tests?

We get anxious about IQ because of the way we understand it. It's just not regarded the same as other kinds of intellectual tests, such as whether you're a good speller, or whether you got a bad grade in calculus. In 1934, one of the leading early researchers on intelligence, Sir Cyril Burt, defined intelligence as "inborn, all-around intellectual ability . . . inherited, not due to teaching or training . . . uninfluenced by industry or zeal [that] enters into all we do or say or think."[1] According to this view, which is not at odds with the way that some psychologists still conceive of intelligence, you are born with a certain amount of intelligence, and there really isn't that much that you can do about changing it. As another pioneer of intelligence testing, Charles Spearman, said about IQ in 1931: "A person can no more be trained to have it in higher degree than he can be trained to be taller."[2] Intelligence is often thought of as largely immutable, the product of innate forces that remain hidden from view, which determines a person's potential to contribute to society. In other words, intelligence is often thought of very much like an *essence*. And as with other essences, the notion of intelligence comes with a lot of psychological baggage.

Does it matter whether people think of intelligence as an essence? It certainly does, and I first learned of some of the consequences of understanding intelligence in essence-like ways when I taught English in Japan. You see, cultures differ in how they think of intelligence, and Japan is one place where people don't think of intelligence as a kind of essence—in contrast to most Westerners. A common experience that I had while teaching in Japan was that I was regularly chastised by the other Japanese teachers for praising the students too much. A student would struggle at providing an answer in broken English and I would encouragingly say, "Good job!" But as the other Japanese teachers told me, "How can you expect them to get better if you're telling them that they're already doing well?" I found this feedback curious, because it was the precise *opposite* of what I had learned about motivation when I studied psychology at a university: I learned that people work harder if they have confidence that they will do well. The reason that we had different attitudes toward praising students is that I thought of intelligence as more like an essence than the other Japanese teachers did.

These disagreements with the Japanese teachers always fascinated me, and I later conducted a cross-cultural study with the psychologist Shinobu Kitayama and others to investigate the powerful role of essentialist beliefs in people's responses to challenges.[3] We had American university students take a kind of creativity test, except that the test was rigged such that it would give them either a really high score or a really low score. For example, students were given three words— *Day*, *Sleep*, and *Fantasy*—and they were asked what one word they could think of that would connect to each of the other three words. Can you think of a word that would work? Give it a try.

Did you come up with the word "dream"? You have *day*dreams, you dream when you *sleep*, and a dream is a *fantasy*. If you thought about it long enough, I bet this answer would come to you because most people in our study came up with it as well. So one group of students in our study got an easy version of the creativity test that consisted only of items like this, which the students tended to get right, and when they graded their own tests, they were informed that their score was better than about 80 percent of the other students from their university. They found out that their creativity skills were really quite impressive. Another group of students wasn't so fortunate. These students got a much harder version of the test. Try to find the word that connects to these three words: *Deal*, *Meal*, *Peg*. Can you come up with it? Keep trying or look to the Notes section to find out the correct answer.[4] Almost no one comes up with the right answer to this one, and the hapless students in this group only saw really difficult items like this. They could only get a few answers right, and when they graded their own tests, they learned that they only did as well as the bottom 20 percent of their fellow students. They learned that their creativity skills were really quite poor. The students were then left alone in a room with a couple of tasks to keep themselves busy while the experimenter was away. One of the tasks was the same kind of creativity task that they had just discovered they were either really good or really bad at. The other task was entirely unrelated; it involved trying to trace shapes without lifting your pencil off the paper. Unbeknownst to our participants, the whole point of our study was to see how long they would persist at each of these tasks while the experimenter was away (when the experimenter returned, the participants were informed of the true purpose of the study and all that had transpired). We also conducted the exact same study in Japan with Japanese university students.

What did we find? Well, those Americans who had learned that they were really good at creativity spent most of their time working on the same creativity task that they had done before. They knew they were good at this task, and they wanted to continue to succeed. In contrast, those Americans who learned that they were bad at creativity instead switched to the line-drawing task. They knew they weren't very good at the creativity task, so they figured they'd be more likely to do well if they tried something else.

In stark contrast, the Japanese students did the precise opposite. Those Japanese who had found out they were good at creativity were more likely to switch to the line-drawing task. Because they were already good at creativity, they might as well try to become good at something else. On the other hand, those Japanese who found out they had done very poorly at the creativity test tried to improve their creativity by sticking to the creativity task. And the key variable that explained this cultural difference was whether people thought of creativity as an essence. The Americans thought of creativity as something that you're born with; the Japanese, in contrast, thought of creativity as the result of hard work. This study shows that one real cost of conceiving of abilities as essences is that people who do so tend to quit in the face of challenges. After all, why make efforts to improve if your ability is ultimately innate and largely immutable? This cost of essential thinking is not at all unique; the psychologist Carol Dweck describes in rich detail a whole host of undesirable consequences that come from thinking of abilities as essence-like.[5] When people think about abilities as grounded in essences, they are more likely to get dejected by failures, have a more negative attitude toward making efforts, be afraid of failure, lie about their own performance, be at risk for depression, avoid pursuing risky options, and try to compete with, rather than cooperate with, members of their own team. There are many pronounced personal and social costs to embracing essence-based theories of abilities.

But just because thinking about abilities as essences can have harmful consequences, it doesn't follow that an essence-based understanding of intelligence must be factually incorrect. What does the evidence have to say about this? Is intelligence just something that we're born with, or do our experiences shape it? To answer these questions, we need to consider the research on the genetics of intelligence.

## Exploring the Genetics of IQ

There are few traits more valued than intelligence, so it's not surprising that there's been a lot of interest in the genetics underlying it. Given the extremely rapid pace of innovations in genetics research, perhaps someday we'll be able to understand the genetics underlying intelligence well enough that we could actually predict someone's intellectual potential merely by looking at his or her genome. One of the world's leading behavioral geneticists, Robert Plomin, recently cowrote a book titled *G is for Genes*, predicting that someday soon we will be measuring intelligence via a gene chip, and then further recommended that schools should customize an educational regimen around the child's unique genetically inscribed talents.[6] Is it really possible to see your intellectual potential written in your DNA?

As a professor, I've always assumed that I must have at least a reasonable IQ, given that I've had to do well on a lot of exams to get a PhD, but I've never found out what my IQ is. Because 23andMe provides information on three genetic markers for intelligence, my genotyping would finally allow me to get some information on my innate capacity for intelligence. I eagerly clicked on the link at 23andMe that provides information on "Measures of Intelligence." It describes how on chromosome 20, at location rs363050, people either have the alleles AA, AG, or GG. Each "A" that you have at this location is associated with 3 extra IQ points. Great, which alleles do I have? GG. Oof! This means I'm down 6 IQ points and I just got started. I turned to the next genetic marker for intelligence provided. Under the link "Effect of breastfeeding on IQ," 23andMe describes research from two studies of two separate SNPs. On chromosome 11, at rs174575, if you have the alleles CC, then you get 6 to 7 extra IQ points if you've been breastfed as an infant. I do have the alleles CC. And also on chromosome 11, at rs1535, if you have the alleles AA, you get 4 to 5 extra IQ points if you were breastfed. I have the alleles AA as well—excellent. But I check with my mother and it turns out that I was *not* breastfed. Oof! It was the late '60s after all, and the conventional wisdom at the time was that breast milk was a poor substitute to all the nutrients that formula provided. So this means I lose *all* of those extra IQ points. Out of the possible 18 IQ points that 23andMe provides information on, I get a grand total of 0. That's a deep hole to be starting from. To get an idea of the magnitude of this difference in IQ, here is what the psychologist Richard Nisbett had to say about the prospects of people who possess IQs of 100 and 118, respectively: "Typical jobs for a person with an IQ of 100 would include skilled worker, office worker with a not terribly responsible job, and salesclerk. A four-year college would have been very difficult for such a person to handle." In contrast, "a person with an IQ of 118 is not only capable of doing excellent college work but is likely to be able to do postgraduate work if that's desired and to become a professional such as a doctor or a lawyer, a high-level manager, or a successful entrepreneur."[7] After receiving this grim news about my poor IQ genes, my own essentialist biases immediately kicked in, and I started to question whether I really was as smart as I used to think I was. Maybe I've just been good at fooling people all along. Your own biases may well be engaged at this moment too, and I wouldn't be surprised if you now are feeling that you can dismiss everything in this book, given that it's the product of such inferior IQ genes.

But let me try to convince both of us that we should be suspicious about 23andMe's IQ genes findings. First, 23andMe itself flagged all of these findings as iffy—they were all given two to three stars, which indicates that the studies are viewed as "preliminary research," in contrast to the four stars given for "established research." None of these studies has ever been replicated, so the validity of these findings is called into question. Second, although 23andMe highlights these three genetic links with IQ, it's important to note that there have been literally dozens of studies investigating the genetics of IQ. And all of this research points to a consensus in the field that there are *no* single genes that predict IQ well. The title of an influential paper on this topic puts it bluntly: "*Most reported genetic associations with general intelligence are probably false positives*"[8] (a false positive indicates a finding that was observed by chance, and is not a genuine finding). A number of extremely large studies that have employed both candidate gene and genome-wide association methods have not found *any* replicable genetic variants that have a large impact on IQ.[9] Arguably, the most optimistic finding yet for identifying genetic predictors of IQ is from a series of investigations with very large samples that found three genetic variants estimated to be responsible for an average of 0.3 IQ points each.[10] So why does 23andMe only provide information from small, unreplicated studies that show unrealistically large estimates of genetic influence on IQ, when all the other research on genes and IQ, employing far more justifiable methods, finds no evidence for these same associations? I think this is another indication that the company is more interested in convincing people that it is able to tell impactful and sensational stories about their lives by looking at their genomes, rather than trying to provide consumers with an accurate picture of the inherent uncertainty in understanding most traits on the basis of their genomes.

So your IQ does not emerge from just a few genetic markers. How many genetic markers are involved in predicting your IQ then? Two very large genome-wide association studies involving thousands of participants have attempted to address this question. The answer: you would need *all* of the SNPs on a gene chip—more than 500,000 of them—to be able to predict half of the individual variation in IQ.[11] Understanding the genetic basis of IQ is a comparable problem to understanding the genetic basis of height.[12] Just as with height, it seems that a reasonable conclusion is that *most genes are IQ genes*. Switch-thinking does not help us to understand intelligence because there are no single IQ genes that have a pronounced impact on our intelligence. Indeed, it seems highly doubtful, contra Robert Plomin, that we'll ever be able to estimate someone's intelligence with much precision merely by looking at his or her genome.[13]

But, just as with height, the fact that there aren't any strong single genetic predictors of IQ doesn't mean that IQ isn't heritable—indeed, it is, just like almost everything else. This means that genes *do* contribute to a person's IQ, even if there are no strong single predictors of IQ in our genes. Two parents with above-average IQ are more likely to bear a child with above-average IQ than are two parents with below-average IQ, even if these children were adopted and raised together in the same environment. But how strong are the contributions from our genes to our IQ?

## How Heritable is IQ?

To answer the question of how heritable IQ really is, we first need to reflect a bit on what heritability means. Recall from Chapter 2 that heritability refers to the percentage of the variance of a trait, such as IQ, in a given population that is due to genetics. Some studies have provided a heritability estimate of IQ as high as .75 to .80,[14] (although these are probably overestimates).[15] A heritability of .80 means that among the individuals tested *within the same population*, about 80 percent of the reasons why some of those people have higher IQ than others is due to their respective genes, and the remaining 20 percent is due to experiences in their lives. And remember that heritability estimates are *always* limited to the sample that they are tested in—they can't speak to people outside of those samples, such as people living in other cultures. Every behavioral geneticist knows this basic fact, but sometimes the conclusions that some draw suggest that even they forget it. It's important to understand this true definition of heritability because heritability estimates vary enormously by sample.

For example, among wealthy populations, heritability estimates of IQ can approach .80, which leads some IQ essentialists to draw conclusions that IQ is largely genetic and that people's experiences have little impact. On the other hand, you get a very different answer when you investigate how heritable IQ is among poorer populations. Heritability estimates of IQ among the poor tend to be much lower than estimates from the wealthy. Although these estimates vary a lot by location, they can approach as low as .10—that is, as little as 10 percent of the variability in IQ scores among poor people may be due to their genes, whereas the remainder (up to 90 percent) is due to people's different experiences.[16] For poor people, genes don't matter as much for accounting for the differences in people's IQ scores. Why is there such a big difference in the magnitude of these estimates, and what does this mean?

These different estimates reveal that heritability does not reflect any absolute contribution of genetics. Rather, heritability within a sample reflects the proportion of influence that comes from genes relative to the proportion of all influences, including both genes and experiences. If people have very similar sets of experiences, then their genetic differences will make up a larger proportion of all the unique influences on them. Consider this thought experiment: Imagine that two human embryos were conceived through in vitro fertilization, were then incubated in identical artificial wombs, delivered, and then raised their whole lives by robots in completely controlled environments that were identical in every possible way—every event of their lives unfolded in the exact same way from the moment of conception. If people's environments could be made identical like this, then pretty much the *only* factor that would influence the different abilities of these two individuals should be their genes. Their heritability estimates for intelligence (and almost everything else) should approach 100 percent because there would be no other aspects of their lives that varied between them. On the other hand, if two people are raised in tremendously different environments from each other, then the influence of their different genes is going to be a fainter cry among the loud din of their different environmental experiences. When environmental experiences are more variable, the estimates of heritability must necessarily be lower.

Higher heritability estimates in wealthier populations rather than poorer ones show that environmental factors that contribute to IQ are more similar among wealthier than poorer populations. Why would this be? Well, a plausible explanation is that among wealthier populations, virtually all children live in a context that challenges them intellectually; they have access to books, have parents and friends who encourage them to take their education at least somewhat seriously, and have rich and stimulating environments that challenge them intellectually. There just isn't all that much variation among wealthy families in terms of having a sufficiently intellectually challenging environment. On the other hand, children from poorer populations have more variation in terms of how much their environments foster their intellectual development.[17] Some children in poor families experience a very rich and stimulating intellectual environment—their parents may encourage them to go the library, invest whatever available income they have toward things like books and tuition, emphasize the importance of their education, and have engaging and challenging conversations with them. On the other hand, some children from poor populations have entirely abysmal educational environments. Their parents may not provide them with access to any books, they may hardly be spoken to, and they may be left to fend for themselves. These very different educational environments will have an impact on children's intellectual development, and because the variability in environment is greater among poorer children, this means that their genetic contribution wields a smaller proportion of influences on their intelligence development.

The social psychologist Richard Nisbett points out that higher heritability IQ estimates among wealthier populations indicates something else—our estimates of the magnitude of heritabilities are likely exaggerated.[18] This is because of the two ways that heritabilities are usually calculated. Some studies compare the similarity of IQ among children to their adopted parents, who share close to 0 percent of unique genetic variants, compared with biological children, who share approximately 50 percent of their genetic variation with their parents. However, there are more families wanting to adopt than there are babies available for adoption, and adoption agencies screen out families that would provide a substandard rearing environment. A consequence of this is that families that adopt are more similar to each other than are other families, and they also provide a better-than-average rearing environment. The average adoptive family scores at about the 70th percentile in terms of a measure of intellectual stimulation.[19] Heritability estimates are thus higher than they would be if the adoptees were raised in families across all social strata.

The second way that heritabilities are estimated is by comparing the similarity of identical twins, who share 100 percent of their genes, to fraternal twins, who only share about 50 percent of their genetic variation. This may seem like an irrelevant point, as the likelihood that someone has a twin has nothing to do with his or her socioeconomic status. However, the likelihood that one shows up for a behavioral genetics study has very much to do with it. In general, people who are attending a university, or who have had university experiences, are both typically more interested in contributing to research, and (especially among current university students) have a much easier time showing up to get involved in such a study. If you are a twin enrolled in a psychology class at a university where behavioral genetics research is taking place, it's highly probable that you will be contacted to participate in a study, and it's quite likely that you will agree to participate. In contrast, if you're a twin living in a poor neighborhood somewhere, it's quite unlikely that behavioral genetics researchers will ever find you, and it's even less likely that you'll end up participating in the study.[20] Regardless of psychology's goals to conduct studies that speak to human nature, they are almost always based on highly biased and nonrepresentative samples that are composed of what my colleagues and I have termed WEIRD (Western, Educated, Industrialized, Rich, and Democratic) populations—a sample that is psychologically unusual in the context of all humans.[21] IQ is heritable, but not to the degree that is usually found in most behavioral genetics studies.

And no matter how accurate estimates of the heritability of IQ are for a given population, it's critical to remember that heritability has nothing to say about whether a trait can be modified.[22] Recall from Chapter 2 that even though the heritability estimates for height approach .90, when Japanese moved from Japan to California in the 1950s, they gained about 5 inches in height.[23] Diet is a key variable that predicts height, and when people in one population have different diets than those in another, it can have a dramatic effect on average height, regardless of the important role of genes in height. Likewise, even though some estimates for the heritability for IQ are as high as .75 to .80, we can see the key role of the environment on IQ when we look at people who change environments. The best way to see this is from adoption studies. When we compare children who are adopted by upper-middle-class parents and those adopted by poorer families, we find that children adopted by upper-middle-class parents have IQ scores that are between 12 and 18 points higher than those adopted by poorer families.[24] This is an enormous difference, and it reveals that our environments have a critical impact on IQ.

Even though heritability has no bearing on whether something can be modified, our essentialist biases cause us to think of anything inherited as an immutable essence. Therefore, anything heritable somehow feels like it should remain impervious to our experiences, even though this is blatantly incorrect. Our essentialist biases transform IQ into a sensitive issue, and it becomes even more uncomfortable when we add ethnic differences into the mix.

## Race, Essences, and IQ

Every year I teach a course on cultural psychology that discusses the ways people's thinking styles differ around the world. Students tend to find it interesting to learn about how, for example, Americans from the South respond more aggressively to insults than those from the North, Japanese look more at the backgrounds when they look at photographs, Russians wallow more in their suffering (yet are less likely to get depressed), Koreans perform worse on cognitive tasks when they try to verbalize their thoughts, and Malaysian men sometimes suffer from an imagined penis-shrinking disorder. The classroom discussions are often animated, and students seem to really enjoy learning about how cultural experiences come to shape the ways that we think. But there is one question that, whenever it is brought up, leaves the class deadly silent and looking uncomfortable. Steven Pinker called it "the most dangerous idea,"[25] and it is often associated with racists. The question is: Does the intelligence of people vary across ethnicities?

One of the most controversial books ever published, *The Bell Curve*, written by Richard Herrnstein and Charles Murray in 1994, argued that it does.[26] Their general argument is thus: IQ predicts people's success at school and the kinds of jobs that they will ultimately get; IQ comes largely from one's genes; on average, Black Americans score somewhat lower on IQ tests than White Americans, who, in turn, score a little lower than Asian Americans; and these racial differences in IQ test performance are probably due to genes. Moreover, Herrnstein and Murray argue that we'll never be able to reduce these racial gaps, given that they're genetic, so we would be better off investing in the education of those who have a high enough IQ to be able to do something with it. As you might expect, with these kinds of contentious conclusions, the book set off an academic firestorm—there have since been far more pages written critiquing their book than there are pages in the actual book itself. I think the book provoked such a sustained angry response, not because it was based on illogical arguments (which would have led it to be ignored), but because Herrnstein and Murray's logic holds up reasonably well *if you accept that IQ is like an essence.* If IQ is indeed like an essence, then the only real way to make sense of any racial differences in IQ would be to conclude that one race has a less intelligent essence than the other. And because IQ is often talked about in essence-like terms, even sometimes by intelligence researchers, Herrnstein's and Murray's basic argument is difficult to dismiss offhand. Herrnstein and Murray cited a survey of more than 1,000 scholars conducted in the 1980s in which only 15 percent said that they believed the racial gap in IQ was due entirely to environmental experiences (that is, genes were not involved). The most common answer offered was that the racial gap in IQ was due to *both* genetic and environmental differences between the races.[27] So the belief that IQ is at least somewhat like an essence that varies between races is not uncommon among many academics.

But, on the other hand, if IQ is *not* like an essence, then Herrnstein's and Murray's logic all falls apart. The racial differences in IQ would instead point to differences in educational experiences, opportunities, and discrimination. So which is the better supported view? Is IQ really like an essence or not? To answer this question, we need to consider three factors: how stable are the differences in IQ across groups, how much of IQ is shaped by experiences, and how the measurement of IQ itself is fundamentally entangled with culture.

## How Stable Are the Differences
## in IQ across Ethnic Groups?

The most controversial argument in *The Bell Curve* was that ethnic differences in IQ performance reflect the different genetic potential of people from different races. But there are several key findings that speak directly against this conclusion. First, we need to remember, as discussed in the last chapter, that race is largely a social construct as opposed to a genetic one—the continental races only explain about 4.3 percent of the genetic variation between people.[28] The vast majority of the genetic differences between people has nothing to do with their ancestral origins, so the very starting point of this argument— assuming that the ethnic difference in IQ must be determined by the genes that distinguish these populations—is on very shaky ground. Second, it's important to note that the magnitude of the IQ gap between Black and White Americans is not a constant; in fact, it has decreased by approximately 5.5 IQ points since 1972;[29] this reveals that the gap is not the product of an unchanging essence, and that as social conditions between the races improve, the gap should be expected to continue to shrink. Third, the magnitude of the racial gap varies enormously by the age of the test-takers. The racial gap between young children is only about 5 points, yet it is closer to 17 points in adulthood[30]—a difference comparable to what's found comparing advantaged to disadvantaged groups in other countries, such as between Maoris and European-descent New Zealanders, between high- and low-caste Indians, or between mainstream Japanese and Burakumin[31] (an outcast group of native Japanese who are confined to stigmatized occupations). That the racial gap increases with age shows that the different experiences that Black and White Americans encounter has an impact on their IQ. To get a quick idea of some of the ways that life experiences differ between races, you only need to look at the sizable differences in the amount of money available to American public schools that are in predominantly Black versus predominantly White neighborhoods.[32]

A fourth piece of evidence comes from the adoption studies we discussed earlier; children adopted by upper-middle-class parents have higher IQ scores than those adopted by poorer families.[33] This suggests that the overall lower socioeconomic status of Black Americans will have a significant causal impact on their IQ. And a fifth piece of evidence comes from research on stereotype threat, which we discussed in Chapter 5. Studies have found that when Black Americans are merely reminded of their race while taking an IQ test, it leads them to perform significantly worse compared with those who are not reminded of their race. It's not hard to imagine how a lifetime of exposure to these kinds of stereotypes, apparently bolstered by scientific arguments such as those advanced in *The Bell Curve*, would lead to increased academic disengagement among Black Americans. All of these findings show that the different kinds of circumstances that Black and White Americans regularly experience have impacts on IQ scores that are larger in magnitude than any of the observed ethnic differences in IQ. This evidence completely refutes the notion that any ethnic differences in IQ reveal something about the genetic potential of the different races.

## How Much of IQ Is Shaped by Experiences?

But the most damning evidence that IQ is not determined by genes comes from one of the more curious findings in the history of intelligence testing. The psychologist James Flynn, working in New Zealand, received a package in the mail in 1984 containing some intelligence test data from the Netherlands.[34] He noted a curious finding: the results of IQ tests from recent years were much higher than those from earlier years. The Dutch national IQ had been increasing! Was this a fluke? Flynn went on to gather the results of IQ tests across time for more than a dozen nations, and by now researchers have amassed data from 31 nations, including developing nations.[35] The same pattern that was found with the Dutch emerges regardless of the country: IQ has been growing around the world! People seem to be getting smarter. This curious pattern has been dubbed "the Flynn Effect," and it reveals just how much our IQ is a function of our experiences.

There are a few things to note about the Flynn Effect that undermine the notion that IQ is like an essence. First, this trend of increasing IQ goes back to the very beginning of IQ testing in the early 20th century, and the rate of increase has been fairly constant.[36] Second, the rates of IQ increases are more pronounced in developing societies than in developed ones, indicating that environmental changes are most impactful when the existing environment is especially lacking.[37] Third, the increases in IQ are not uniform across all different types of intelligence. There have been very small, almost nonexistent, increases in vocabulary, arithmetic, and information. The gains have been considerably larger for other kinds of measures of intelligence, such as comprehension, spatial IQ, and reasoning. And the gains are largest in one particular measure of IQ, known as Raven's Progressive Matrices. How large are the IQ gains on the Raven's test? Well, from 1947 to the 2002, a period of approximately two generations, the average increase in IQ was approximately 27.5 points. This is a massive gain, and is almost the same magnitude as the difference between someone of average IQ (IQ tests are normalized to fix the average IQ of a population at 100) and someone who would be designated as having an intellectual disability (formerly called mental retardation; the cutoff for this designation is usually set at 70). The Raven's test is viewed as a measure of fluid intelligence—that is, the kind of intelligence that allows you to solve a novel problem on the spot, with virtually no prior knowledge. This is in contrast to crystallized intelligence, which is the kind of intelligence that relies on stored knowledge, such as your knowledge of the names of the capitals of countries around the world. The Raven's test relies on abstract reasoning ability; the items show a series of different geometric shapes, and people are asked to choose the next shape that would complete a pattern (see the example below—the answer is in the Notes section).[38]

## Which of the following shapes should replace the question mark?

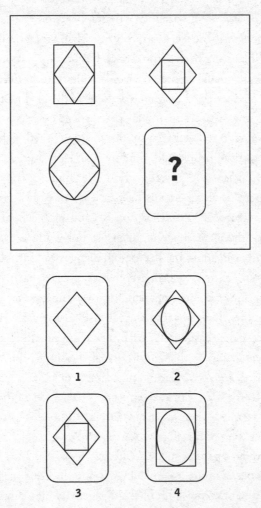

These different rates of increase highlight how our brains develop in response to our experiences, and these differences suggest that we are now living in a world that provides us with more experience to develop certain kinds of intellectual skills more than others. For example, our vocabulary and arithmetic skills have not improved much over time because we are not living in a more demanding world in these respects. Indeed, people tend to read books less and watch television more now than they did at the time of the Second World War, and the typical level of vocabulary that appears on television is only about the fourth-grade level.[39] In contrast, abstract reasoning abilities have been increasing because our world has become a more challenging place and demands more problem-solving from us than before.

How can we see that our world has come to demand more problem-solving? First, we can see that today people need more education than they did before to succeed at most jobs. In the United States in 1890, only 27 percent of people had more than an eighth-grade education, and only 9 percent had any education after high school, whereas in 2012, 94 percent of Americans have at least an eighth-grade education, and 58 percent have education beyond high school.[40] Further evidence for the increasing complexity in our world is that we seem to be seeking out more intellectual challenges for fun. The science writer Steven Johnson makes a convincing case that popular culture has become far more challenging over the past few decades.[41] We can see this change in terms of television programs having more complicated plots than before, with traversing timelines and dozens of characters (compare the complexity of the single, linear storyline of *Starsky and Hutch* with the tangled knot of plotlines of *Game of Thrones* or *Lost*). Even more extreme are the changes in the complexity of video games. There were few abstract problems to solve in the early video games of the '80s, such as *Space Invaders* or *Donkey Kong*. However, today's games, such as *Civilization V*, *Starcraft II*, or massively multiplayer online role-playing games, immerse the player in a constant series of complex problems to solve. By living in an increasingly complex and challenging world, we learn the kinds of abstract problem-solving skills necessary to thrive in it. And the resulting rise in the average person's IQ in recent decades dwarfs the magnitude of any identified differences between ethnic groups.

## The Measurement of IQ Is Forever Entangled with Culture

It is especially noteworthy that the largest gains in IQ over time are found on the Raven's Matrices test because this test was designed specifically to be a "culture-free" IQ test—that is, it was believed that your cultural experiences would not affect how well you did on the test. Why would the culture of an IQ test matter? To get an idea, see how you do on the following IQ items, which were part of IQ testing of the US Army in the First World War.[42]

1. The Percheron is a kind of
   *(a) goat, (b) horse, (c) cow, (d) sheep.*
2. The most prominent industry of Gloucester is
   *(a) fishing, (b) packing, (c) brewing, (d) automobiles.*
3. "There's a reason" is an advertisement for a
   *(a) drink, (b) revolver, (c) flour, (d) cleanser.*
4. The Knight engine is used in the
   *(a) Packard, (b) Stearns, (c) Lozier, (d) Pierce Arrow.*
5. The stanchion is used in
   *(a) fishing, (b) hunting, (c) farming, (d) motoring.*

Oh, did you have a hard time on these? Well, perhaps you'd have known more of the answers if you had a better set of IQ genes. That is the conclusion that was usually drawn for those army recruits who struggled on these tests. But obviously these questions require sufficient exposure to certain cultural knowledge, and without that knowledge, these items become entirely inappropriate for assessing someone's intelligence. The Raven's IQ test was designed to precisely avoid this kind of problem.

Yet it turns out that even the Raven's test depends on cultural knowledge. To realize this, we have to look at people who have entirely different cultural experiences from those of us who live in modern, industrialized societies. Several studies have assessed various cognitive skills among people living in small-scale indigenous societies. In these societies, people acquire their food, shelter, and tools directly from the environment around them, in ways largely similar to those of their ancestors thousands of years ago. And these studies with indigenous populations all lead to the same conclusion: intellectual skills are the product of our experiences.

For example, the anthropologist Helen Davis gave the Raven's test to the Tsimane, a foraging horticultural people who live deep in the rain forests of Bolivia. How did they do on the test? Well, to put it into context, "intellectually average" 11-year-olds get 78 percent or more of the questions correct, and "intellectually defective" (the outdated label that the test manual used for very poor performance) 11-year-olds got at least 47 percent correct.[43] So what percent correct did Tsimane 11-year-olds get? Just 31 percent—a score that is so deep within the "intellectually defective" range that it is completely off of the reference charts.[44] However, another group of the Tsimane who lived in villages that had schools was also tested, and this group fared much better. When tested as adults, those Tsimane who had received the limited education that these schools provided scored 72 percent correct, compared with the unschooled Tsimane adults, who scored just like the unschooled children, getting only 31 percent correct.

These results show two key things. First, this is clear evidence that the Raven's is not in any way a culture-free test, because those Tsimane who live in villages that have schools get more than twice as many IQ items correct as those who live in villages without schools (and this is even though the schools are rather rudimentary in their offerings). Formal educational experiences had a massive impact on the Tsimanes' ability to engage in abstract reasoning. Second, if we could take their IQ scores as a direct reflection of their intellectual potential, then the Tsimane who live in villages without schools would appear to be severely "intellectually defective." If we took these IQ scores at face value, we'd have to conclude that the whole community should be completely unable to take care of itself, and the entire tribe would need to be institutionalized. Of course, this would be a ridiculous conclusion to draw. If you met them, the Tsimane would never strike you as people who were not intelligent. They have phenomenal tracking abilities, are able to make all their tools and provision themselves from the forest around them, and have a vast knowledge about many hundreds of species of flora and fauna. They have been surviving for centuries in an extremely demanding environment. That is, the cognitive skills of the Tsimane have developed to master the challenges that their environment places on them, and the Raven's test simply does not tap into those skills. It's not a reflection of some kind of true universal intelligence; it just reflects how well they can answer those items.

It's not just measures of IQ that are forever entangled in cultural experiences. We get similar kinds of findings with some indigenous populations for very different kinds of cognitive skills, including some remarkably simple ones. Take, for example, some research that was conducted with the Pirahã, a hunting and gathering tribe in the Brazilian Amazon.[45] The Pirahã do not have words for numbers in their language beyond words for "around one," "around two," and "many," and they receive no formal training in how to count. Like the Tsimane, they show amazing foraging and tracking skills that enable them to thrive in such a challenging environment as the Amazon. However, despite this obvious evidence of their intelligence, adult Pirahã do not have the quantitative skills of the average American 4-year-old. Here's what happened in one experiment: An adult Pirahã is shown an empty can and he watches as an experimenter places six nuts into it, one after the other. Next, he watches the experimenter take five nuts out, one at a time. The question is then posed to him: Are there any nuts remaining in the can? The Pirahã man has no idea—he can't count or keep track of specific quantities. Without the cultural knowledge of familiarity with numbers, the Pirahã lack what, by Western standards, would seem to be the most rudimentary of cognitive skills.

Here's an even more extreme example: some adult members of the Hadza, a hunting and gathering tribe in Tanzania, were given some simple six-piece jigsaw puzzles to do—the same kind of puzzles that 4-year-old American children do all the time. The Hadza had never seen puzzles before, and they were largely stumped by them—most couldn't solve them at all, and those who could took several minutes to complete them.[46] The Hadza adults struggled on the kind of task that American preschoolers do for fun, and they struggled despite that they are known as the "the most cognitively complex foragers on Earth,"[47] and have an incredibly rich set of intellectual skills to allow them to thrive in their very demanding ecosystem.

These examples highlight how intelligence is not an innate skill but rather emerges from what we learn from our environments. Had you been raised without schooling in an environment like that of the Tsimane, Pirahã, or Hadza, you too couldn't solve quite simple items on the Raven's test, you wouldn't be able to count nuts in a can, and you'd find a children's jigsaw puzzle to be about the hardest problem you faced that day. On the other hand, you'd have mastered a wealth of information and skills necessary to survive in some of the more ecologically challenging environments on the planet—skills that few readers of this book would possess. I imagine if I were left to fend for myself in the wilds of Tanzania, I probably would die of thirst or starvation, if I hadn't already succumbed from eating some poisonous berries or from ending up in a hyena's den. We come to learn intellectual skills that allow us to confront the demands from our own environments, and our brains change over the course of our lives in response to our experiences.

In sum, intelligence is nothing like an essence. But because of our essentialist biases, we intuitively expect that group differences in any domain, including intelligence, are likely the result of group differences in genes, and we overlook the ways that people's experiences affect their intelligence. Herrnstein and Murray couldn't resist the attraction to this faulty essence-based intuition, and neither can many others who have offered similar racist theories for intelligence.

In the eugenics movement in the early 20th century, the notion that IQ was an essence was also wrapped up with concerns about the essence of another human quality: a penchant toward criminality. The key concern of the eugenics movement was that the so-called "feeble-minded"—a widely used term that captured both those who did poorly in school and those who had criminal inclinations—were going to outbreed those of sound mind and slowly "contaminate" the human race.

## Criminal Genes

The 1954 Broadway play *Bad Seed* tells the story of a doting and self-sacrificing mother named Christine and her most delightful and charming 8-year-old daughter, Rhoda. But it soon becomes clear that Rhoda isn't as innocent as she first appears. Once, Rhoda was jealous to learn that her friend Claude had won a "good penmanship" medal that she herself had wanted. So she went to see Claude at a dock on a lake, removed her tap-dancing shoes, and bludgeoned him with the steel tips of her shoes until he surrendered his medal. And then, for good measure, Rhoda rolled Claude off the dock into the lake and hammered his fingers with her shoes each time he tried to grab the dock to pull himself up, until he ultimately drowned. She similarly dispatched some other people whose existence was inconvenient to her, all while presenting herself as the most adorable little girl imaginable. How could Rhoda be capable of such unspeakable horrors?

The play provides a tidy answer: Rhoda's mother, Christine, had been adopted as a young child and never got to know her own biological mother, who, it turns out, had been a serial murderer herself. Apparently, Christine's mother's homicidal maniac genes somehow skipped a generation, and her sinister essence had again become evil incarnate in little, adorable Rhoda. As the play put it, some people "are bad seeds—just plain bad from the beginning, and nothing can change them."[48]

This play resonates with a quest that dates back centuries. What is the best way to make sense of why people sometimes commit heinous acts of violence? The most horrible acts seem so beyond the pale that we feel a natural urge to focus on the essence of the offender. The most satisfying answer for why Rhoda committed her monstrous crimes is that there must be something about Rhoda that is rotten to the core; it was her evil essence that caused her vicious acts.

The 19th-century British psychiatrist Henry Maudsley put it simply: "the wicked are not wicked by deliberate choice . . . but by an inclination of their natures." Such a worldview has broad implications—if the root of evil lies within one's nature, and not as the product of one's experiences, then how can we ever work toward preventing criminal acts? Maudsley went on to further describe criminality as an "intractable malady. . . . The dog returns to its vomit and the sow to its wallowing in the mire. . . . How can that which has been forming through generations be re-formed within the term of a single life? Can the Ethiopian change his skin or the leopard his spots?"[49] As Maudsley saw it, the problem with criminals is their essence makes it impossible for them to avoid a life of crime.

As extreme as Maudsley's views were, they weren't uncommon for his era. The incipient field of criminology was just being launched at this time by the 19th-century Italian criminologist Cesare Lombroso, the most significant figure in the history of criminology.[50] Lombroso maintained that whereas most criminal acts were committed opportunistically by people in unfortunate situations, about 40 percent of criminals simply had no choice but to do evil; they were born that way. Their underlying essence predisposed them toward a life of crime. Lombroso spent much of his career trying to find ways to detect that underlying criminal essence. He felt it was discernible by carefully studying the head and face, using calipers to measure various aspects of the skull and carefully documenting various facial cues that betrayed the inner malady. For example, he described how rapists "nearly always have sparkling eyes, delicate features, and swollen lips and eyelids. Most of them are frail, some are hunchbacked."[51] If one was a careful observer and had some calipers at the ready, he or she could detect who was an innate criminal before any crime was committed. Lombroso's theory thus transformed the question of understanding a criminal from *what one had done* to *who one was.*[52] And according to this view, there would be no hope in rehabilitating a born criminal; the only solutions were either lifelong incarceration or capital punishment.

Much has changed since Lombroso's time, and researchers are no longer searching for a criminal essence by measuring whether one has swollen lips or an unusually shaped skull.[53] But in some respects, the idea of an innate criminal has been preserved—it's just that the tools of the genomics age appear more sophisticated than Lombroso's calipers.

There have been a variety of attempts to find individual genes that might point to a criminal nature. The first clue that a specific gene could be involved came from a Dutch extended family with a broad history of criminal behavior. One of the men in this family raped his sister, and then after being sent to prison for this, stabbed a prison warden in the chest with a pitchfork. Another male in the family tried to run over his boss with a car, and yet another forced his sister to undress at knifepoint. Two other men in the family were known arsonists, and many of them had recorded cases of exhibitionism and voyeurism. On the other hand, there were also many men in the family who had no prior history of violence.[54] In 1993 researchers found that the men with the criminal history could be distinguished from the men with the spotless records in the makeup of one particular gene: monoamine oxidase A (MAOA). The violent men had a very rare mutation that effectively silenced the expression of this gene, which resulted in a buildup of various neurotransmitters in their brain that caused greater hyperactivity and poor impulse control. Although this particular mutation is extremely rare, and is noteworthy in how devastating its effects are on the neurochemistry of the brain, there are other, more common variants along this same gene that have also been implicated in violent behavior.

The most famous study to explore the relation of the MAOA gene with violence was conducted with men in New Zealand, where researchers looked at two common variants of this gene. Specifically, the researchers targeted the interaction between whether one had a low-expressing or a high-expressing variant of this gene and whether one had experienced much abuse as a child.[55] Among those who had little past history of abuse, the particular MAOA variants that they possessed had no impact on their behavior. In contrast, among those who had been the victim of severe child abuse, the rates of antisocial behavior were higher among those with the variant for low expression of MAOA compared with those with the variant for high expression. The toxic combination of a particular genetic variant and a history of child abuse led to a greater likelihood of violence. Findings like these have led some to label the MAOA gene "the warrior gene," which gives the impression that it is a key cause, and perhaps the ultimate cause, of violence. Even more problematic, since the low-expressing variants of this gene are not spread equally around the world, some have speculated that different frequencies of the MAOA alleles could help explain ethnic differences in violence. For example, Maoris, famous for their warrior past, have been found to be more likely to have the supposedly problematic MAOA variants compared with people of European ancestry, as have African Americans.[56] Perhaps these differences in allele frequencies could explain why both Maoris and African Americans are overrepresented in prison populations in New Zealand and the United States.[57]

However, there are some important reasons why the so-called warrior gene is not the reason behind any ethnic differences in crime. First, the relations between this gene and a history of violence are not particularly strong or consistent, they don't hold up at all for women, and even these weak effects for men are dependent on a history of severe child abuse. Moreover, one follow-up study in the United States found that this interaction between child abuse and the MAOA gene only emerged for White males; there was no such inter-action among non-White males.[58] So perhaps it's more appropriate to think of MAOA as "the White male warrior gene."

However, any labels like "the warrior gene" are highly problem-atic because they suggest that this gene is specifically associated with violence. It's not, just as the alleles from other genes do not only have one outcome. Pleiotropy is the term for how a single genetic variant can influence multiple different phenotypes. MAOA is highly pleio-tropic: the traits and conditions potentially connected to the MAOA gene include Alzheimer's, anorexia, autism, body mass index, bone mineral density, chronic fatigue syndrome, depression, extraversion, hypertension, individualism, insomnia, intelligence, memory, neu-roticism, obesity, openness to experience, persistence, restless leg syn-drome, schizophrenia, social phobia, sudden infant death syndrome, time perception, and voting behavior.[59] Perhaps it would be more fitting to call MAOA "the everything but the kitchen sink gene."

In addition, focusing on the parallels between the MAOA gene and the higher rates of imprisonment among Maoris and African Americans relative to White New Zealanders and White Ameri-cans ignores the MAOA allele frequencies of the rest of the world. The low-expressing variants of this gene are not unique to Maoris and African Americans but are also more pronounced among many other societies that are not known for being particularly violent. For example, Japan has the lowest homicide rate of any large nation on the planet,[60] yet Japanese are more likely than Europeans, African Americans, and Maori to have the low-expressing MAOA alleles.[61] However intuitively satisfying it may be to try to explain cultural differences in violence in terms of genes, as of yet there is no direct evidence for this.

## Thinking about Criminality Genes

Although the evidence for genetic predictors of violence remains rather spotty at present, it's still useful to ask what would happen if research ever did uncover strong evidence for a genetic basis of criminality. How might we respond to the idea that criminality was bound up in our genes? Would this knowledge influence the ways that people are prosecuted for crimes?

One possible response to locating the cause of criminality in our genes is that it could lead juries to become more sympathetic to the accused. A guiding principle of legal decision making is *mens rea*, which literally translates as a guilty mind. If genes were perceived to be the ultimate cause, then it could be argued that the accused couldn't control himself or herself, and thus lacked *mens rea*. Indeed, versions of this "my genes made me do it" defense have been used in hundreds of court trials with varying success. The very first attempt of a genetic defense came shortly after research on the Dutch family with the problematic MAOA genes was published. In 1994, Stephen Mobley was being tried for an execution-style murder of a pizza store manager in the state of Georgia. His lawyers petitioned to have Mobley genotyped to see whether he had the same genetic mutation as the Dutch family. They reasoned that if he did indeed possess the mutation, it could be seen as a mitigating factor that might save Mobley from a death sentence. In this case, the judge denied the request, concluding that the evidence for this particular genetic link was not yet sufficient, so Mobley was never genotyped, and he was ultimately executed in 2005. However, other judges have responded differently to genetic arguments. In 2009, an appeal hearing in Italy reviewed the sentence that had been given to Abdelmalek Bayout, an Algerian who confessed to murdering Walter Felipe Novoa Perez because Perez had insulted Bayout over the eye makeup that he had been wearing. A psychiatric report indicated that Bayout had the low-expressing variant of the MAOA gene that previous research had linked to higher rates of violence when combined with a history of child abuse. Upon considering this evidence, the judge acknowledged that he found the research on the MAOA gene to be compelling, and he shortened Bayout's sentence by an additional year.[62]

The judge in Bayout's case does not appear to be an anomaly. Researchers in one psychological experiment investigated the reactions of almost 200 US state trial judges to genetic information that was included in a hypothetical case.[63] The case was modeled after Stephen Mobley's murder of the pizza store manager. One-half of the judges just read a description of the accused's situation, whereas the other half of the judges read the identical information but also were told about research on the MAOA allele and violence, and were informed that the accused did in fact have the problematic allele. Just as in the Bayout trial, the judges who learned of the relation between the MAOA gene and violence assigned a sentence that was on average one year shorter than those judges who didn't hear about the genetic story. The judges who learned about the research linking the MAOA gene to violence in some situations were more likely to argue that the defendant lacked control for his actions.

Of course it's possible that the judges who read about the MAOA research were more lenient to the defendant simply because they had more information available to them than those judges who didn't read about the genetic research. Perhaps just encountering *any* kind of causal explanation of violence would lead to a reduced perception of causality. My student Benjamin Cheung and I investigated this question by collecting the reactions of Americans adults to different causes of violence.[64] All the participants read a fictitious story about a college student named Patrick who had fatally stabbed another man after an altercation. The description of the crime was identical for all participants, with the exception that some participants read some additional information about scientific research on causes of violence. One-third of the participants read that the accused had an allele that research had shown was associated with violence; specifically, they were informed that those who possessed the problematic allele were four times more likely than people with other alleles to engage in violent behavior. Another third of the participants read that the accused had been severely abused as a child, and that research had found that people who had been abused as children were also four times more likely to engage in violent behavior. And the final group of participants did not read about any research about potential causes of violence. Then, participants were asked whether they felt that it was reasonable for Patrick's lawyer to make a defense claim that Patrick was not fully responsible by reason of insanity. Those who read that Patrick possessed a gene that predisposed him to violence were more than twice as likely to view an insanity defense as applicable than those were who read about Patrick's history of abuse. Moreover, people who read that Patrick had problematic genes were also more likely to explicitly say that his actions were less under his control than those who read about his history of abuse.[65] People seem to be willing to forgive Patrick for his genes but not for his upbringing, and this is because we think our genes make up our essences.

But is someone really any less responsible for their actions if his or her genes are implicated? A problem with this argument is that we would be hard-pressed to find any actions that we engage in where our genes are not involved—our behaviors do not occur in any gene-free zones. Or, consider this: there actually is a particular genetic variant that, if you possess it, makes you about 40 times more likely to engage in same-sex homicide than those who possess a different variant.[66] It's known as the Y chromosome—that is, people who possess it are biologically male. Given this, should we infer that Y chromosomes cause murders, and thus give a reduced sentence to anyone who is the carrier of such a chromosome because he is not really responsible for his actions? The philosopher Stephen Morse calls the tendency to excuse a crime because of a biological cause the "fundamental psycholegal error."[67] The problem with this tendency is that it involves separating your genes from yourself. Saying "my genes made me do it" doesn't make sense because there is no "I" that is independent of your genetic makeup. But curiously, once genes are implicated, people seem to feel that the accused is no longer fully in control of his or her actions.

Genes might seem to exonerate bad behavior, but it's important to remember that the double-edged sword of essentialism can cut both ways. Not only does thinking about the genetics of criminality lead people to view the accused as less responsible for his or her crimes, but it can also lead people to be *more afraid* of the accused. After all, if someone's genes are seen as the ultimate cause of his or her crime, and that person still has those same genes, then what is he or she going to do in the future? Genetic accounts should lead people to expect the accused to offend again.

Take the case of Gary Cossey—a captain of a New York rescue squad. In 2005, he was arrested for being in possession of child pornography. Typically, this crime is associated with a sentence of approximately 4.5 years. But in Cossey's case, Judge Gary Sharpe issued a sentence of 6.5 years because he predicted that within fifty years, scientists would prove that child pornography was caused by "a gene that you were born with," and "not a gene you can get rid of." Judge Sharpe further declared that "you are what you're born with. And that's the only explanation for what I see here. . . . It is something you could not control."[68] Just as in Bayout's murder trial, this judge also saw problematic genes as the ultimate cause of Cossey's crime. In this case, however, rather than being more lenient in the sentencing because of any lack of personal responsibility, Cossey was to be punished more because his problematic genes would cause him to continue to reoffend throughout his life. An appeals court later overturned Sharpe's ruling for basing a conclusion on imagined future scientific results, and Cossey was sentenced more leniently by another judge.

We've found that many people resonate with the sentiment expressed by Judge Sharpe. In the study discussed earlier in which we presented participants with both genetic and environmental accounts for Patrick's murderous impulse, we found evidence for both sides of the double-edged sword. Participants who read about a genetic account for Patrick's violence felt not only that he was less responsible for his actions but also that he was more likely to reoffend in the future. When people think about genes underlying criminality, they can simultaneously feel both sympathy that the accused is not fully responsible for his crimes and a desire to lock him up and throw away the key before he follows his inner nature and offends again.

## When Essences Meet Eugenics

We've seen there can be real costs to our genetic essentialist biases: thinking about "criminal genes" can undermine our beliefs in individual responsibility and rehabilitation. Thinking about "intelligence genes" makes us more likely to quit in the face of difficulty and more pessimistic about investing in education. But surely the most pronounced cost of essentialist thinking has to be what unfolded in the first half of the 20th century, with the birth of the eugenics movement. The rationale underlying this movement was straightforward: if essences are what make a person intellectually dull or have a criminal nature, then perhaps society should play an active role in cultivating more desirable essences by deciding who is allowed to reproduce. Just as a cattle rancher strives to mate his best bull with his most productive cow, the eugenics movement sought selective breeding among humans.

Eugenics, which the journalist Christine Kenneally calls "the worst idea in history,"[69] seems so obviously sinister now that you would be hard-pressed to find a sane person today willing to publicly endorse it. After all, hundreds of thousands were forcefully sterilized and millions of people were murdered in Nazi Germany under its banner for concerns that they possessed the "wrong" essence to pass on to future generations, whether that be a Jewish essence, a gay essence, a Roma essence, or run-of-the-mill "feeble-mindedness." Because the Nazis promulgated such an abhorrent strain of eugenics, it is tempting to link the whole movement with dangerous and crackpot ideologues. But the Nazis certainly were not unique in embracing a eugenic worldview.

One could see the burgeoning eugenics movement on full display in the United States by roaming the stalls of the 1924 Kansas Free Fair. Alongside the displays of the local farmers' giant pumpkins, plump hogs, and prized Holsteins, one would encounter the bounty of some of the region's other great crops: the Jenkins, the Coopers, and the Schmidts. The fair held "Fitter Family" contests in which families would be subjected to batteries of measurements, medical inspections, physical challenges, and IQ tests. After these tests were finished and the results tabulated, the winning families would receive medals, along with the coveted designation by the Kansas governor of being recognized as "Grade A Individuals." The larger the families, the more prestigious the medals they won, as the ultimate goal of these contests was to encourage those people with "the right stuff" to reproduce their top-notch genomes in service of their nation.[70]

The Fitter Family contests are an example of "positive eugenics"— positive in the sense that the goal was to increase desirable genes in the country's collective "germ plasm." However, it wasn't long before these efforts were supplanted by much more extensive programs of negative eugenics—efforts to weed out those who were seen to carry an inferior essence. Negative eugenics gained much traction in the United States in 1927. At the time, the Virginia State Colony for

Epileptics and Feeble-Minded sought to sterilize one of its wards, Carrie Buck, after she gave birth to an illegitimate child. Carrie was an 18-year-old ne'er-do-well with a supposed mental age of 9, and because she didn't want to go through with the sterilization, the case was ultimately heard by the Supreme Court. The case for sterilizing Carrie hinged on her apparent feeble-mindedness being seen as hereditary. Carrie's own mother, Emma, had a record of prostitution and immorality, and reportedly had a mental age of only 8 years old. And following a cursory examination of Carrie's 7-month-old daughter, Vivian, it was concluded that she too had "a look" that was "not quite normal."[71] Supreme Court Justice Oliver Wendell Holmes Jr. infamously wrote for the majority, using language that still shocks to this day: "It is better for all the world if, instead of waiting to execute degenerate offspring for crime, or to let them starve for their imbecility, society can prevent those who are manifestly unfit from continuing their kind. . . . Three generations of imbeciles is enough."[72] Following this landmark ruling, the number of forced sterilizations in the United States skyrocketed, and over the next several years, more than 60,000 Americans, disproportionately minorities and women, were sterilized.[73] Similar mandatory sterilization regimes emerged in such diverse places as Canada, Sweden, Japan, and much of Latin America, and these remained in place in several countries until the late 1970s.[74] Whether or not one had a child was no longer just the decision of the parents involved—the state had the ultimate say in who would be allowed to reproduce.

Popular support for eugenics didn't emerge from a vacuum but rather grew out of impassioned discussions among leading public intellectuals of the time. Indeed, the list of prominent eugenics supporters in the early 20th century reads like a volume of *Who's Who*. Alexander Graham Bell was the chairman of the board of scientific advisers to the Eugenics Record Office. Eugenic ideas were popularized in novels by H. G. Wells and plays by George Bernard Shaw.[75] The eugenics movement received generous funding from the likes of the families of Rockefeller, Carnegie, Guggenheim, Eastman, Kellogg, and Vanderbilt.[76] University presidents, such as David Starr Jordan from Stanford and A. Lawrence Lowell from Harvard, were key supporters, and by 1928, there were courses on eugenics being offered in 376 different universities.[77] Teddy Roosevelt gave several speeches urging the right kind of people (that is, Anglo-Saxon Americans of good stock) to fulfill their patriotic duty of reproducing for their country, and he is credited as having done more to popularize eugenic ideas than any other individual.[78] Although the movement was most popular in the United States among White Protestants, its support even extended to some African American intellectuals, such as W. E. B. DuBois, who believed that only "fit" Blacks should procreate in order to improve the image of the race.[79] Likewise, the National Association for the Advancement of Colored People promoted eugenics by hosting "Better Baby" contests, sharing the same goal as that of White eugenicists to improve the genetic stock of the country, with the proceeds of these contests going to its antilynching campaign.[80]

In the early 20th century, eugenics was embraced by politicians of all stripes, and in contrast to most political issues, its support did not fall neatly along any left–right ideological divide. Conservatives were attracted to eugenics by the thought of increasing the relative proportion of their own kind, whereas liberals recognized that a welfare state could only succeed if there were fewer needy people that were dependent on it. The politician who introduced universal health care to Canada, Tommy Douglas—the former premier of Saskatchewan, father of actor Donald Sutherland, and the man voted in a popular television show as "The Greatest Canadian"—wrote in favor of eugenics in his Master's thesis in 1933, arguing that the country's social ills would also need to be addressed with biological solutions.[81] Likewise, the first sterilization law in Europe was passed in Denmark, followed by other Nordic countries, just as their own social welfare states were being launched.[82] Progressives at the time imagined eugenic utopias, where a combination of sterilization laws and welfare offerings would ultimately reduce the amount of human suffering.[83]

You can also find some curious eugenic roots in a number of contemporary organizations where you might least expect to find them. For example, although Planned Parenthood today is associated with women's reproductive rights, its original founder, Margaret Sanger, was a vocal eugenicist. She famously stated, "More children from the fit, less from the unfit—that is the chief issue of birth control." The field of genetic counseling (formerly termed "genetic hygiene," emphasizing its focus on genetic purity), in which people are advised of their risks for passing down any potential inherited disorders, originally emerged out of an attempt to prevent unfit babies from being born. Although the field today emphasizes that it now offers advice to allow parents to make their own decisions, rather than providing more coercive directive counseling, some still criticize the endeavor for having eugenic goals.[84] And two of the largest environmentalist organizations, the Sierra Club and the Save the Redwoods League, also embraced some key eugenic tenets at the time of their founding. The Redwoods League viewed the stately majesty of the redwood as a symbol for a great race of humans, and the Sierra Club recognized that environmental conservation would ultimately depend on population control.[85] Given that these organizations were born in a time when racist sentiments of "inferior" immigrants were commonly held, they viewed it as best for the planet if certain people didn't reproduce.

I note with considerable shame the deep eugenic roots in my own field of psychology. Because eugenicists strived to create a system where only those of sufficient genetic quality would be allowed to reproduce, they needed a way to measure one's overall genetic value. Psychologists stepped up to this challenge by developing ways to measure intelligence, which served as a proxy for one's genetic potential, and this pursuit was a key concern of the field prior to

World War II. Although it's certainly not emphasized in any of our major textbooks, many of the founding fathers of psychology were active members in the eugenics movement, including Carl Brigham, James McKeen Cattell, Robert Fisher, G. Stanley Hall, Karl Pearson, Charles Spearman, Lewis Terman, Edward Thorndike, and Robert Yerkes.[86] Intelligence tests, like the kinds described earlier in this chapter, became a key tool in the eugenics movement for identifying both who should be allowed to immigrate into the United States, and ultimately to determine who should be sterilized.

But even though eugenic supporters were found in all corners of society (especially among the well-to-do), the field most closely linked with eugenics prior to World War II was that of genetics. In the early days of genetics, there was little distinction between being a geneticist and being a eugenicist. Francis Galton, seen as both the father of behavioral genetics and eugenics, illustrates the common origins of these two fields. Likewise, it's noteworthy that every member of the founding editorial board of the journal *Genetics*, established in 1916, endorsed the eugenics movement.[87] The connection between genetics and eugenics was also clearly seen in the first professional genetics society—the American Breeders Association, established in 1903, which was primarily concerned with improving the lines of both livestock and humans.[88] Prior to the war, eugenics was simply thought of as applied human genetics—the fields were related in the same way as chemistry and chemical engineering.[89] This connection also held in Germany—more than half of academic biologists joined the Nazi Party, the largest representation of any professional group.[90]

Although eugenics soon became verboten following World War II, this was mostly because people were so horrified about the actions of the Nazis, not because of a widespread understanding of the scientific limitations of eugenic ideas.[91] Hence, following the war eugenic ideas went largely underground, with strategic name changes to avoid the negative connotations of the past. The journals *Eugenics Quarterly* and *Annals of Eugenics* were renamed *Social Biology* and *Annals of Human Genetics*, respectively. Likewise, The American Eugenics Society and The Eugenics Society were rechristened The Society for the Study of Social Biology and The Galton Institute, respectively.[92] And every now and then you may hear a famous geneticist making statements that sound awfully eugenic in their tone. For example, Francis Crick, the codiscoverer of DNA, stated, "I do not see why people should have the right to have children." He proposed that a licensing scheme might be appropriate, whereby "if the parents were genetically unfavorable, they might be allowed to have only one child, or possibly two under special circumstances."[93] Likewise, the other codiscoverer of DNA, James Watson, has argued that stupidity should be thought of as a disease, and that his solution to deal with poverty would be to get rid of the bottom 10 percent of the IQ bell curve.[94]

An important reason why the fields of human genetics and eugenics were so closely linked in the early days of genetics research was because of the simple switch-thinking stories about heritability that were common at the time. Complex human traits were often assumed to emerge in the ways uncovered by Mendel—just as it only took one gene to make a pea plant grow yellow seeds, most human traits were also assumed to emerge in a similar fashion. For example, a leading American eugenicist, Charles B. Davenport, proposed that a love of the sea came from a single sex-linked gene, for how else could one explain that sea captains were usually male and the sons of sea captains themselves?[95] Davenport also proposed simple switch-thinking accounts for a series of rather ridiculous conditions that he came up with such as "nomadism," "shiftlessness," and the "innate eroticism" that afflicted "wayward women."[96] And, the bogeyman of the era, feeble-mindedness, was also thought to be "a condition of mind or brain which is transmitted as regularly and surely as color of hair or eyes."[97] If this was the case, it followed that if you could somehow just keep carriers of the feeble-mindedness gene from breeding, you could rid the world of the condition.* For example, in the textbook *Genetics*, the biologist Herbert E. Walter argued that if you could enforce a sterilization law across the whole United States, "less than four generations could eliminate 9/10 of the crime, insanity, and sickness of the present generation in our land."[98]

I and my students Cermet Ream and Benjamin Cheung were curious whether these same kinds of links between deterministic genetic accounts and support for eugenics existed in the present, even though eugenics is no longer viewed as a viable ideology. To explore this, we gave large samples of Canadian university students and American adults a scale that tapped into support for eugenics. This included items such as "There should be laws discouraging people with low intelligence from having biological children," and "Anyone convicted of a violent crime should be sterilized as part of their punishment." We were happy to discover that most people disagreed with these eugenic items, although there was quite a range of opinions. Then we also gave people some measures of beliefs in genetic determinism, including such items as "I am of the opinion that intelligence is a trait that is strongly determined by genetic predispositions," and "I believe that an analysis of my genetic predispositions will allow a trained scientist to predict many of my abilities and traits without having any personal knowledge of me."[99] We found that those who believed that genes were deterministic were far more likely to support

---

* Actually, some eugenicists, such as Henry H. Goddard, assumed that feeble-mindedness was a recessive trait. The problem with this logic is that it should have argued *against* eugenic efforts, as was noted by some critics at the time. This is because removing those with feeble-mindedness from the gene pool would have only removed a small fraction of those who were carriers of the trait (they would need to be homozygous to have the phenotype), and the prevalence of the trait would thus have hardly been affected. For more discussion of this point, see Paul, D. B. (1995). *Controlling human heredity: 1865 to the present* (p. 68). New York, NY: Humanity Books.

eugenics than those who didn't view genes that way.[100] Thinking of genes as the main cause of life outcomes seems to grease the path toward eugenic ideology.

Because essentialist views of genes are still widely shared, the allure of eugenics has never really disappeared. However, as the sociologist Troy Duster argues, this time around, eugenics isn't entering through the front door, as it did before the war. Rather, eugenic ideologies are now emerging from the back door, through various treatments, screenings, and therapies that are part of the new science of genetic engineering.[101]

# 8.

# A Brave New World: Engineering Better Essences

**IN 2009, THE CHONGQING CHILDREN'S PALACE** in China made an offer straight out of *Brave New World*. A group of children aged 3 to 12 years old were taking part in a new program that employed DNA testing to identify genetic gifts. A saliva swab and a gene chip were used to identify 11 different genes that provided information on the child's IQ, emotional control, height, memory, listening ability, focus, athletic ability, and more. The Children's Palace then used these results to recommend to parents which careers would be best for their children to pursue. For example, after looking at the results of one child's 11 genes, Dr. Huang Xinhua, a leading scientist on the project, told CNN, "This child is very thoughtful and focused, so I suggest she go into management."[1] The parents of these children paid $880 for the test, with the hope of giving their children a leg up on the world by identifying their natural talents from the get-go. Why pay to have your child spend a few

years at a university trying to figure out what he or she wants to do
in life when the answer can be read directly from his or her genome?
Chen Zhongyan, the mother of a four-year-old girl in the program,
said, "It's better to develop her talents earlier rather than later. Now
we can find when she is young, and raise her based on what her natural gifts are." News of the plans of the Children's Palace led to articles
around the world questioning whether other countries were going to
fall behind China as it maximized the potential of each of its more
than one billion citizens.

But lest you're leery of China taking over the world through its
genetic sorting, you can be reassured that some Western companies
have aspired to do the same. For example, in 2009, the American
company My Gene Profile offered to provide a career plan for your
child based on the results of 40 different genes that were examined.
Likewise, in 2008, Atlas Sports Genetics promised to help you find
out what sports your child is most likely to succeed in. The company's
marketing pitch stated: "Finding any great Olympic champion nor-
mally takes years to determine. What if we knew a part of the answer
when we were born?" Although many of these direct-to-consumer
companies have ultimately gone bankrupt, they continue to prolifer-
ate, and their numbers almost doubled from 2013 to 2015.[2]

There is something deeply seductive about the promises of these
companies, though. I imagine if you were told that these companies
had tested your genome, it would be hard to resist the urge to see
what they had found out. And the results might just lead you to question some of your life decisions. "Did I make the right career choice?
Perhaps I should have tried to become an engineer instead." Or, "I
knew I should have tried out for the high school track team! Maybe
it's not too late to try to pick up my genetically destined sport now?"

Before we get ahead of ourselves, however, I should point out that
right now the science is nowhere near the hype—this idea remains an
armchair fantasy. None of the genes identified in these tests can offer
much precision at predicting any of the traits that direct-to-consumer

companies purport they do. The companies making these promises have been criticized for using genetic testing "to sell new versions of snake oil,"[3] and some of them have received cease-and-desist letters from the FDA.[4] Consider their claims about a single gene predicting your intelligence with any degree of precision. As discussed in Chapter 7, such a claim is a complete fiction, as the most promising single genetic variant associated with IQ predicts only 0.3 IQ points. Likewise, the most discussed athletics gene, ACTN3, which direct-to-consumer companies often rely on to predict whether you should be a sprinter or an endurance runner, is similarly limited. The so-called sprinting genotype, which consists of two copies of the R allele, is found in only 50 percent of elite sprinters compared with 30 percent of nonathletes, and the endurance genotype (two copies of the X allele of the same gene) is found in only 24 percent of elite endurance runners, compared with 18 percent of nonathletes.[5] Most Olympic-class sprinters and endurance runners do just fine in their sports without the ideal ACTN3 variants; this shouldn't be surprising, given that the ACTN3 variants only account for 2 percent to 3 percent of muscle function.[6] A consensus statement published by a team of geneticists concluded that "the information provided by DTC [direct-to-consumer companies] is virtually meaningless for prediction and/or optimisation of sport performance."[7] And it's worth noting that these tests are all ultimately measuring the wrong quality; people turn to these tests because they want to know about their child's phenotype, such as whether their child will be intelligent or a good sprinter. You're going to get a far more accurate answer to this question by ignoring genotypes entirely, and instead measuring phenotypes directly. The best way to assess how smart or fast your child will be is to give him or her an IQ test or time how long it takes him or her to run to the end of the track.

The reason we're so captivated by the false promises of these direct-to-consumer companies is that they're a perfect fit with our essentialist biases. Hearing that your child has "the sprinter gene" makes it sound like the child is a born sprinter, even if this just means that, holding everything else constant, your child's muscles should only be 2 percent to 3 percent faster than those of the average person. It doesn't seem to really matter to us that most genetic links to phenotypes are incredibly weak; our attraction to essence-based explanations leads us to forget that both our intelligence and our running skills are massively predicted by our educational experiences and physical training, respectively.

But even though our future potential can't currently be read with any precision by looking at a small handful of common genetic variants, we are presently sitting at the cusp of a genetic engineering revolution, and now is the time to consider the ethical concerns future genetic technology will raise. How should we be thinking about genetic engineering? And how will our essentialist biases distort our understanding of these exciting but disturbing new technologies?

## Genetically Modified Organisms and You

Is there a more disturbing scientific idea than genetic engineering? We can see just how bothersome this notion is by looking at the ways that science fiction stories so often deal with it; books and movies typically discuss genetic engineering in dark dystopian undertones, where our humanity has been lost in favor of some short-sighted chasing of scientific progress, such as in *Gattaca*, *Brave New World*, or *Jurassic World*. In fact, genetic engineering troubles us so much that our fear and discomfort extends to species far removed from humans.

One of the most politically contentious scientific debates concerns the use of genetically modified organisms (GMOs) in agriculture. GMOs are created when novel genes are directly injected into the cells of other organisms, most often into plant cells. Believe it or not, one of the most widely used GMO technologies is derived from a naturally occurring event—genes can hitch a ride with a common soil bacterium called *Agrobacterium tumefasciens*. Until recently, *Agrobacterium* had only been a pest for farmers, as it causes tumors to grow in plants. A few decades ago, scientists discovered that *Agrobacterium* causes these tumors by inserting foreign genes into plants, and soon after it was realized that this bacterium could become a useful vector for inserting desired genes into plants.[8] An alternative method for inserting novel genes into a plant is to use a specially modified BB gun that literally fires microscopic gene-coated gold or tungsten bullets directly into a plant's cells. Using both of these methods, the inserted genes become incorporated into the organism's own DNA, and the genes get expressed along with all of the other preexisting genes in the organism.[9] These methods allow for a nearly infinite number of novel proteins to be added to any new GMOs.

The most common reason for creating a novel genetically modified (GM) food is to introduce a gene that helps solve a particular problem faced by the organism. For example, in response to the epidemic of the deadly papaya ringspot virus, the gene from the SunUp papaya was introduced to the Kapoto varietal, which resulted in a papaya that was resistant to the virus.[10] There is as of yet no other method to control the ringspot virus, and today 80 percent of Hawaiian papaya is genetically engineered.[11] Researchers have also explored ways to protect tomatoes from frost damage. One potential solution was discovered: the *afa3* gene from the winter flounder, which is known as "the antifreeze gene," plays a key role in preventing ice crystals from forming in the blood of this fish while it swims in the cold ocean waters of northern Canada. If this same gene could be transferred to tomatoes, perhaps they could be grown in colder climates. Researchers successfully transferred the *afa3* gene to tomatoes, where it produced the same protein in the tomato that it did in the fish.[12] However, this protein didn't have the desired effect of protecting tomatoes from frost, and it has never been made commercially available.[13]

Another example comes from pigs. Because pigs are so physiologically and genetically similar to humans, they are often studied as a model organism for human disease. The main difficulty with doing research on pigs, however, is that they are so large; the most commonly used pig for research, the Bama pig, weighs around 100 pounds. To address this problem, researchers have disabled the growth receptor gene in the pigs' fetal cells, which resulted in a pig that grows to only one-third of its typical size. The small pigs have proven so popular that they are now being sold as pets—the first breed of GM pets.[14]

Alternatively, rather than using genetic engineering to correct a shortcoming in an organism, it can also be used to modify existing species to create entirely novel benefits. For example, a new species of rice, "golden" rice, was modified by inserting genes from a daffodil that allowed the rice to produce beta carotene, a precursor to vitamin A.[15] Golden rice tastes just like normal rice, and aside from the presence of beta carotene and its yellow hue, it is identical to normal rice in every way. If golden rice were to replace the normal rice that people usually eat, it could ultimately protect up to a half-million children who go blind each year from vitamin A deficiency.[16]

These kinds of GMOs may sound as though they're straight out of science fiction, but you've certainly already had firsthand experience with GM food products. A striking amount of agricultural products, particularly those grown in the United States, have been genetically modified. For example, more than 80 percent of corn and more than 90 percent of soybeans grown in the United States are GM products. As much as 80 percent of American processed food contains at least some GM content,[17] although this typically is not evident from looking at the labels.

If you find it disturbing to think that you've been consuming plants that have had their genes deliberately altered by scientists, you're not alone. Fears of the consequences of consuming GM foods—often called Frankenfoods by their critics—are widespread. A 2003 survey found that more than half of Americans felt that "GM food threatens the natural order of things," and about two thirds felt that "serious accidents involving GM foods are bound to happen."[18] This distrust has only been increasing: a 2013 *New York Times* poll found a full three-quarters of Americans expressed concern about GMOs in their food, with the most common worry being perceived health risks.[19] In particular, people are especially concerned about the idea of GM animals being created,[20] which became a more prominent concern in 2015 when the FDA approved the first GM animal for human consumption, a salmon that grows to maturity in half the time. And attitudes toward GM products are even more critical among Europeans than they are among Americans.[21]

A widespread and growing fear about GMOs suggests that there might be something harmful about them that people are slowly coming to understand. So which people are most fearful of GMO products? I and my students Cermet Ream and Benjamin Cheung investigated this by asking Canadian and American adults some questions about GM foods, such as "How risky would you say GM foods are in terms of risk to human health?" or "How willing would you be to purchase genetically modified foods if it was more nutritious than similar food that isn't genetically modified?" We found that people's answers to these questions were clearly predicted by one thing—their knowledge about genes. The higher their genetic literacy, the *less* fearful they were of GM products.[22] Unfortunately, genetic literacy is scarce. For example, one study found that only 57 percent of Americans and 36 percent of Europeans were aware that "ordinary tomatoes contain genes."[23] Another common misconception about GMOs, endorsed by 20 percent of Europeans, is that a person's own genes can be modified by consuming GM food.[24] And people seem to be aware that they don't know much about the process of creating GMOs—54 percent of American adults admitted that they knew very little or nothing at all about biotechnology, yet they still are deeply bothered by it.[25] The contrast in attitudes toward GMOs is most striking between scientists and the general public; whereas only 37 percent of American adults are open to GM food, a full 88 percent of scientists embrace the technology. In fact, there is even more of a consensus among scientists that GMOs are safe than there is among scientists that humans are contributing to climate change (87 percent).[26] The American Association for the Advancement of Science has gone so far as to conclude that "every . . . respected organization that has examined the evidence has come to the same conclusion: consuming foods containing ingredients derived from GM crops is no riskier than consuming the same foods containing ingredients from crop plants modified by conventional plant improvement techniques."[27] Nonetheless, most of the general public is staunchly against the idea of consuming GM products.

A key reason why scientists aren't overly worried about GMOs is that they do not generally see much of a distinction between GM agricultural products and other kinds of agricultural products. The truth is, genes have been greatly modified in all of the agricultural products that we consume, not only through GMO technology, but through simple domestication. For example, wheat was one of the very first crops grown by humans, and archaeologists have noted the genetic modifications in the seeds of wheat from the very first years that it was cultivated in present-day Israel, approximately 10,000 years ago. Those stalks of wheat with seeds that were more likely to remain attached to the stalk, because of a sturdier spikelet, were more likely to be selected by the earliest farmers to plant. Likewise, those stalks with bigger seeds, and a greater number of seeds, were also selected, and over the years the wheat that was cultivated evolved through this process of artificial selection so that it became an entirely different species from that which was growing nearby in the wild.[28] A similar story of genetic changes through domestication could be told for all of the major crops that are consumed around the world.

Over time, these efforts to change the nature of our crops and livestock have become more conscious, directed, and technically sophisticated. Varieties of one plant may be cross-bred with those of another, resulting in entirely new varietals that bear properties from each of the parent plants; similar cross-breeding efforts have led to the more than 100 different breeds of dogs around the world. And since the 1950s, plant breeders have been using radiation from gamma rays and X-rays on seeds with the goal of hastening the rate of mutations in the seeds, and then later seeing which of these mutations led to beneficial effects.[29] All of these technologies create a wide variety of genetic modifications, and not all of them are desirable.[30] The case with dog breeding shows this clearly—many dog breeds suffer from genetic diseases, such as congenital deafness in Dalmatians or mitral valve disease in Cavalier King Charles Spaniels. The key difference between these other kinds of breeding efforts and GMOs is that genetic engineering technologies can ensure that only the specified genes are modified, while leaving the rest of the genome intact.

There are other aspects of GMO production, however, that are bothersome to many. The patents of GMOs are owned by large agri-businesses, such as Monsanto, Dupont, and Syngenta, who have complete control over the distribution of these products. I doubt I'm the only person wary of such centralized control over our food supply; it would be natural for these companies to put their own interests ahead of the consumer, and to keep trade secrets about the nature of their crops from both farmers and consumers. Moreover, this centralization will lead to a general homogenization of our food supply, with fewer varieties of crops being grown. And because many of the GMO crops are produced to be resistant to other herbicides, such as Monsanto's Roundup, we run the risk of our crops being doused with increasingly large amounts of these herbicides, with potential ill health effects for both the consumer and the environment.[31] Furthermore, this rapid and expansive transformation in the way that agriculture is conducted could also potentially lead to far-reaching changes in the ecosystem. Critics of GMOs have a long list of concerns about the practice that bother them.

Yet it's not the monopolization of our food supply or the soaking of our environment with frightening chemicals that alarm most people when it comes to GMOs. Rather, most opponents of GMOs are against the technology because it violates what psychologists call an "absolute moral value."[32] Absolute moral values differ from other kinds of moral values in that people tend to hold on to them no matter the consequences—regardless of the costs or benefits, violating an absolute moral value feels deeply wrong.[33] For example, most people view incest between brothers and sisters as an absolute moral violation. You probably share this value too. When asked why they are against sibling incest, most people are able to come up with some reasons, such as they are worried that incest might lead to babies with genetic abnormalities or that the sexual relations might wreck the family. But researchers have found that reasons such as these don't seem to be the *real* reasons why people are against incest. For example, research participants have been asked, "What if the brother and sister used birth control so there wouldn't be any babies?" Or "what if they said that their sexual experience deepened their commitment to each other? If these were true, would you become open to the idea about sibling incest?" For most people, these questions don't change their views toward incest[34]—they still view it as just plain wrong and disgusting, no matter the costs or benefits.

And this seems to be the case for most people's attitudes toward GMOs as well. It doesn't seem to matter to them whether the products might help people in developing countries, whether they would reduce the use of artificial pesticides, or if they are distributed by nonprofit organizations.[35] About 70 percent of those who are opposed to GMOs are opposed as an absolute moral value—they claimed that GMOs should be prohibited, no matter what the risks and benefits were, and regardless if everyone else thought the products were harmless.[36]

GMOs lead to moral outrage because the very idea violates several of our essentialist biases. First, we tend to think of essences as being natural; we think of our food as coming directly from God in all its natural goodness, straight from the Garden of Eden, brimming with purity and nutrition. Conversely, all that is not natural seems to be potentially dangerous—as mentioned in Chapter 3, this is the naturalistic fallacy. For example, research finds that the more natural that people perceive a food product to be, the more they are inclined to prefer it,[37] and the more upset they are about the idea that the food contains GM products.[38] Yet there are numerous examples that expose this fallacy. Cassava, better known to Americans as tapioca, is one of the most widely consumed starches in the tropics, yet its preparation involves a number of time-consuming steps to extract the poisonous cyanide that exists naturally within it. Wild almonds are also rich in cyanide, and it's only through various domestication efforts that a genetic mutation arose that led to the cyanide-free variant that is now grown in orchards around the world. Naturally existing fatal poisons exist in many of the foods that we eat—rhubarb leaves contain oxalate; cashews, urushiol; green potatoes, solanine; nutmeg, myristicin—and almost every year in Japan people die from eating pufferfish that has not had all of its tetrodotoxin properly removed. But tellingly, the toxins that exist in natural plants don't seem to attract the same fears as the imagined toxins in GM products.

Yet modifying the genes of a plant can seem like a violation of the rules of nature, and the scientists who engage in this are often perceived to be, in the words of Prince Charles, "tak[ing] mankind into realms that belong to God and God alone." This is a commonly voiced fear in the GMO debates, and in people's fears about genetics more generally.[39] For example, one of the most vocal critics of GM crops, Vandana Shiva, who condemned the Indian government for accepting the GM food aid that the United States sent to India following a 1999 cyclone, states "GMO stands for 'God, Move Over,' we are the creators now."[40] Scientists who participate in creating GMOs, it follows, seem to have made a Faustian bargain with the whole world lying at stake. These fears aren't uncommon—a full two-thirds of American adults don't trust scientists on the issue about the safety of GMOs.[41] This lack of trust for scientists conducting GMO research is revealing because generally, scientists belong to one of the most respected occupations.[42] Clearly, science dealing with genetic modification triggers deeper fears than other scientific fields.

A second way that GMOs get tangled up with our essentialist biases is that they cross the boundaries of essences. There are two general kinds of gene transformations conducted in GMO research. *Cisgenic* transformations involve the introduction of a gene from a plant that is either of the same species or is one that can be naturally cross-bred, such as introducing a gene from a Fuji apple to a Golden Delicious apple. *Transgenic* transformations involve the introduction of a gene from a species that cannot be naturally cross-bred, such as a gene from the winter flounder being introduced into a tomato. From the perspective of geneticists, it doesn't matter where the gene comes from; it matters what the gene does—that is, what protein it codes for. Remember that humans share the same genes with many other species; we have genes that are also in cockroaches, yeast, winter flounders, and tomatoes, and this reflects that humans have many proteins that are similar in kind to those from these other species. However, in our essentialist minds, this distinction between cisgenic and transgenic transformations is key. We think each species possesses a distinct essence, and genes are thought of as the carriers of those essences. A fish gene, for example, is seen to represent the whole essence of the fish, with its distinctive smell, shape, and slipperiness. When a gene is perceived to cross the border that divides species and essences, it appears to violate the categories that we depend on to make sense of our worlds. A survey conducted in 24 different countries backed this up; it found that people from all countries viewed transgenic transformations as more disturbing, more unnatural, riskier, less useful, and more likely to harm the environment, compared with cisgenic ones.[43] An example of the ways that people fear and misunderstand transgenic transformations can be seen in a survey that investigated whether people thought a GM tomato with a fish gene would "taste fishy." Only 42 percent correctly answered that it would not.[44] Many people instead seemed to feel as if they were responding to a recipe that combined tomato and fish essences; such a tomato might taste something like an anchovy pizza.

Indeed, a tomato with a fish gene sounds to many like an abomination. And even though this GM tomato never made it out of the lab, it's become a powerful rallying cry for the anti-GMO movement, with the image of a tomato with eyes and fins frequently appearing on protest posters. Opponents of GM foods have tried to emphasize the many perceived risks of these products, often pointing to purported studies that reveal how GM food products are rife with dangers to our health.[45] The problem is that most of these studies either don't support the claims made by critics of genetic modification, don't exist, have never been published, or have been retracted.[46] For example, in 2012, a paper was published purporting that rats fed genetically modified corn were far more likely to develop a variety of tumors, pituitary gland problems, and liver and kidney problems.[47] The paper was championed by GMO critics and continues to be frequently cited, despite being retracted in 2014 because serious problems with the experimental design undermined all of the findings.[48] If there were legitimate concerns about the health risks of GM foods, then we'd expect to see many published studies in peer-reviewed journals that documented them—Monsanto does not have a monopoly on funding scientific research on GM foods. And the best evidence that GM products are safe for our health is that they've already been in our food supply for the past couple of decades with no obvious ill effects. We fear GM products more than we should because these products lie at odds with our essentialist biases.

## Engineering the Family

If the genetic engineering of our food supply is bothersome to the majority of the industrialized world, it's not surprising that an even more fearful idea is the genetic engineering of humans. Until recently, mating decisions were the only way to engineer your family. In the past half century, however, a series of rapid technological advancements has given us new ways to shape our family's genetic makeup. Some of these technologies are available already and are being used by parents to select the genetic makeup of their children, while others are still being developed, and their possible applicability to humans lies in the not-too-distant future.

### *Eliminating Some Genetic Diseases*

The most commonly used genetic engineering on humans aims to reduce or eliminate the presence of certain genetic diseases. A number of different technological advances have made it possible to identify, in the womb, whether fetuses have any major chromosomal abnormalities, such as possessing extra chromosomes or having missing chromosomes, or are missing large sections within their chromosomes. Some examples of these include Down syndrome, caused by an extra chromosome 21; Turner syndrome, in which girls have one X chromosome but are missing the other sex chromosome; and Angelman syndrome, in which a segment on chromosome 15 is missing. Because chromosomal disorders involve the alteration of large regions of the genome that contain multiple genes, their effects can often be broad and devastating to the development of the fetus.

The first technology to emerge that allowed an investigation of the chromosomes prior to birth was amniocentesis. Researchers discovered that one could retrieve fetal cells from the amniotic fluid in a mother's womb, and in 1968, this was first applied to diagnosing Down syndrome in utero. A significant limitation of amniocentesis, however, is that it cannot be performed reliably until the 14th to 16th weeks of pregnancy, and it is an invasive enough procedure that it is associated with a 0.5 percent to 1 percent miscarriage rate.[49] In 2011, though, a new technique for genotyping fetuses was developed. Because the DNA in a pregnant mother's blood contains both the mother's DNA as well as DNA from the fetus (2 percent to 6 percent of the DNA in a mother's blood comes from the fetus), it has become possible to isolate the fetus's DNA and screen it for a wide variety of different conditions. This "cell-free fetal DNA screening" has become especially popular because it can be performed as early as 10 weeks into gestation, is highly accurate, and unlike procedures such as amniocentesis, poses no risk to the fetus.[50]

Although it remains a contentious ethical issue, especially among those on the religious right, it is not uncommon for mothers who discover that their baby has a serious genetic disorder to seek an abortion. For example, prenatal diagnoses of Down syndrome, the most common chromosomal abnormality, lead to abortions about 92 percent of the time in Europe,[51] and approximately 60 percent to 80 percent of the time in the United States.[52] Moreover, abortion rates of babies with Down syndrome have increased by as much as 34 percent since the advent of cell-free fetal DNA screening. All of these procedures represent a kind of genetic engineering because they reduce the proportion of babies born with various chromosomal disorders.

But chromosomal disorders are not the only kinds of serious genetic diseases that can be identified in utero. Although the vast majority of cases of genetic diseases involve many genes, there are more than 15,000 Mendelian genes that have been identified in the database kept at the Online Mendelian Inheritance in Man,[53] and many of these, though rare, are associated with diseases with devastating or even fatal consequences. One category of Mendelian diseases is recessive; that is, the disease phenotype only emerges if you inherit the recessive allele from both of your parents. Chances are that you are a carrier of at least one recessive allele that is linked to a Mendelian disease, which means that if you reproduce with someone who is a carrier for the same condition, your child would have a one-in-four chance of having both copies of the recessive allele and would develop the disorder. For example, 23andMe identifies me as a carrier of recessive alleles associated with both phenylketonuria and alpha-1 antitrypsin deficiency—two disorders with devastating health outcomes.* This means that if my partner was also a carrier of either of these, then our child would have a one-in-four chance of developing these disorders. What can be done to reduce the number of children born with these kinds of Mendelian diseases?

Concerns with recessive genetic disorders are especially pronounced in some insular communities, where these disorders are more commonplace, likely as a result of genetic drift. For example, among Ashkenazi Jews, a number of recessive diseases are more common than in the population at large, including Tay-Sachs, Canavan disease, and cystic fibrosis. These higher frequencies are especially problematic in the Orthodox Jewish community because people are selecting mates from a small, insular population with relatively high risk for these genetic diseases. In an effort to reduce the number of cases of these diseases, in 1983, Rabbi Joseph Ekstein came up with a solution: he created Dor Yorshorim, Hebrew for "upright generation," and also known as the Committee for Prevention of Jewish Genetic Diseases.[54] The committee strives to reach a balance between reducing the number of cases of these genetic diseases without stigmatizing

---

* Interestingly, whereas 23andMe identifies me as a carrier of the recessive genes associated with phenylketonuria and alpha-1 antitrypsin deficiency, the whole exome sequencing done on me by Gentle identified me as the carrier for alpha-1 antitrypsin deficiency and Meckel syndrome. The marker for Meckel syndrome is not tested within 23andMe's gene chip; however, both 23andMe and Gentle agree that I'm a carrier of the same SNP, rs5030860, that 23andMe links to phenylketonuria. The difference is that Gentle informs me that the symptoms associated with this SNP do not fit the presentation of "classic phenylketonuria." Hence, even the evidence for seemingly straightforward cases of recessive disorders is not always agreed upon.

people who are carriers of any of the problematic alleles. Here's how it works: a Dor Yorshorim representative visits schools and recruits students to be tested for their carrier status for a variety of recessive conditions. The students only receive a personal ID number which is not linked to their names in any way. Later, when the student is older and interested in getting married, he or she calls the Dor Yorshorim number and provides the personal ID numbers for both partners in the prospective marriage. A return call comes after approximately eight minutes. In most cases, the person would be informed that the partners are compatible (that is, they are not both carriers for any of the same genetic diseases that are tested for). On the other hand, if both partners are carriers for the same recessive disorder, they would then receive genetic counseling informing them that they would have a one-in four-chance of any children developing a life-threatening disorder; the vast majority opts to select a different partner upon getting this news. This way, people do not learn of their carrier status unless they are planning on marrying someone who is also a carrier for the same condition; otherwise, the secrets of their genetic information remain locked in an anonymous vault. This eugenic solution has reduced the number of children born with Tay-Sachs in the Orthodox community in North America from 50 to 60 a year in the 1980s to 4 to 6 a year in the 2000s.[55]

Since the advent of direct-to-consumer genomics companies, you don't need to participate in Dor Yorshorim to be able to learn about your and your partner's genetic risks. For example, you could discover that you and your partner share carrier status for a recessive disorder after comparing your 23andMe results. What do you do then? Are you willing to roll the dice and stay with this partner and hope that your children beat the odds of the genetic lottery, or do you decide to just not have any biological children of your own? If none of these options sound desirable, a third option has recently emerged: preimplantation genetic diagnosis (PGD).

PGD is performed on embryos that are created through in vitro fertilization (IVF) procedures. Sperm is introduced to fertilize egg cells in vitro, and the resulting embryos are allowed to grow for three days until they have developed into an eight-celled stage. At this point in development, the embryos consist solely of eight identical undifferentiated stem cells, prior to any of the cells developing into their more specialized roles, such as blood cells, muscle cells, or neurons. One of these eight cells can be nudged off from the embryo, and the missing cell will be quickly replicated, returning the embryo to its original eight-celled stage. The cell that has been separated from the embryo can then be genotyped, making it possible to identify whether it is free of the harmful gene in question. The embryos that are created using PGD represent something like a set of parallel universes—each of them could develop into your future children, but some of them have genes that will lead them certainly to contract a disease, whereas others do not. Only those embryos that are free of the harmful recessive genes would be selected to be implanted back in the mother's womb. Using this technology, you could change the odds of your child being born with a known recessive disorder from 25 percent to essentially zero, or from being born with a known autosomal dominant disorder like Huntington's from 50 percent to zero. PGD could largely eliminate any Mendelian diseases from couples who know of their genetic risks beforehand. And not only would a couple's children be free of these genetic diseases themselves, but their entire line of descendants would also be spared from being a carrier of these diseases.

An advantage that PGD has over other genetic screening technologies discussed earlier, such as amniocentesis or cell-free fetal DNA screening, is that no abortions are involved. Rather than terminating a fetus with an identified genetic abnormality, PGD allows one to choose which embryos are to be implanted in the mother's womb to begin with. Based on the potential benefits of PGD in reducing the number of people born with disabling conditions and diseases, some bioethicists have argued that it is a moral failure to not use PGD.

For example, Jacob M. Appel has argued for mandatory screening of all embryos at IVF clinics.[56] Likewise, Julian Savulescu argues that parents are "morally required" to use PGD to genetically engineer the best offspring that they can on the grounds that we should all strive to provide our children with the best opportunities that are available.[57] These views by some bioethicists have stirred up much controversy, and PGD is still fraught with ethical implications. As with other genetic engineering technologies, PGD is often perceived as deeply unnatural, and the most frequently voiced concern is that the technology has people playing the role of God.[58] Another concern is that PGD stigmatizes disabilities. For example, the ethicist David Wasserman notes that using PGD to avoid disabling diseases "appears to reflect the judgment that lives with disabilities are so burdensome to the disabled child, her family, and society that their avoidance is a health care priority."[59] Likewise, one of the early PGD pioneers, the obstetrician turned bioethicist Jeffrey Nisker, argued that if we use PGD to "strive for perfection, we are going to blame people with disabilities. We're not going to accommodate them, or support them with tax dollars."[60] Despite these criticisms, a majority of Americans favor the use of PGD in at least some circumstances. A survey conducted with American adults in 2013 found that 73 percent were in favor of using PGD to screen for genetic diseases that are fatal in the first few years of life, with considerably less support, 48 percent, for screening of diseases that may not occur until later in life.[61] These numbers showed little change from a similar survey conducted over a decade earlier, when far fewer people knew about PGD.[62] For most Americans, avoiding a fatal childhood disease is seen as an acceptable tradeoff for choosing PGD, despite how unnatural or God-like it may seem.

The ethical implications of PGD get even more challenging when we consider uses beyond the screening of genetic diseases. Consider the case of Molly Nash.[63] She was born on July 4, 1994, but as soon as she was delivered it was clear that all was not well. Molly had the rare and devastating recessive disorder Fanconia anemia, which left her without thumbs or hip sockets, and a somewhat malformed brain and heart. But most threatening of all was that her condition also put her at high risk for developing leukemia before she was of school age. To survive, she needed a bone marrow donor who shared her same set of leukocyte antigen genes. (The success rate for bone marrow transplants is low when donors are not immunological matches.) But any of her future siblings would only have a one-in four-chance of being a good donor match. Facing these discouraging odds, Molly's parents decided to use PGD to select an embryo for Molly's future brother that did not share the same Fanconia anemia disorder as Molly had, but did carry the same set of human leukocyte antigen genes as she did. Her brother was a made-to-order bone marrow donor, and his parents gave him the name "Adam," reminiscent of God using the rib of Adam to create Eve. Unlike almost all babies born before him, Adam was born to serve as a means to an end; he was created with PGD so that his bone marrow could be used to save his sister's life. Adam and Molly Nash's story served as the inspiration for the 2009 movie *My Sister's Keeper*, which poignantly wrestled with the ethics of creating a baby to be used as spare parts for another sibling.

The ethical implications of PGD get even more troubling when we consider cases beyond the good intentions of these "savior siblings" and consider the kinds of traits that might be selected for using PGD.[64] To strengthen their connection with their children, some parents have tried to ensure that their children share the same disability as they do, such as deaf parents wanting to have a deaf baby, or dwarf parents wanting their child to also be a dwarf—conditions which can sometimes be caused by a single gene.[65] In one survey, four American IVF clinics reported that they had provided PGD to select embryos that were chosen to have certain disabilities.[66] And one of the most common reasons for using PGD has become "family balancing"—that is, people use PGD to select the sex of their future child. A full 80 percent of American fertility clinics allow PGD to be used for sex selection,[67] although this stands at odds with popular opinion: only 21 percent of American adults are in support of PGD being used for sex selection.[68]

Because PGD allows one to peruse the complete genome of any embryo, the technology can potentially be used to select for *any* kind of genes, not just ones related to diseases or sex. In principle, it could be used to select for traits such as eye color or specific athletic skills. The dystopian science fiction film *Gattaca* foresaw a future when PGD was used for precisely these purposes. But there's a huge practical limitation of PGD: the universe of options for one's future children is limited to the kinds of embryos that a couple can produce. If neither parent has any genes for, say, blue eyes, then PGD would not help create a blue-eyed baby. Or if there were a specific combination of multiple genes that a couple wanted to select, it is highly unlikely that any embryo the couple created would possess that exact combination. To actually enhance your children's genomes beyond the possibilities you can create with your partner, other technologies are necessary.

### Genetic Enhancements

Is it possible to have offspring with traits that are better than your own? Perhaps this could be seen as the ultimate kind of parental love—a desire to provide your children with a better life by giving them a better set of genes.[69] What kinds of options are there for parents who want to enhance the genes of their children?

First, we should note that people have always strived to get their children the best possible genes that they can through the evolutionary mechanism of "romantic attraction." That is, people are generally attracted to those who have the most desirable traits—those who are attractive, intelligent, and caring, people with charming personalities and athletic skills—and these traits are all predicted by people's genes. By the logic of evolution, those of our ancestors who were most attracted to partners with desirable traits, and were able to mate with them, were more likely to have children with those same desirable traits, which would enhance their children's own survival and reproductive success. We can see evidence of our evolved preferences in the simple fact that there is actually considerable agreement about who the most attractive and desirable mates are. For example, when we see the contestants at the Miss Universe pageant, or see *Elle Magazine*'s listing of the most eligible bachelors, we generally are in agreement that we are looking at a rather desirable set of potential mates. For the most part—one exception is that we are also attracted to people who share our own idiosyncratic backgrounds and interests[70]—there's a fair degree of consensus on who the 10s are out there, and we all find those 10s to be attractive. The problem is there just aren't enough of them to go around. "The matching hypothesis,"[71] which has to be one of the most obvious theories to come out of psychology, states that we are likely to end up with a partner who is roughly at our own level of "mate value," to use the cold terminology of evolutionary reasoning. So while we may all be attracted to the 10s, they tend to get paired up with the other 10s; then we start to look at the 9s, but they usually end up with the other 9s, and so on until we get to the point where we are considering partners who are roughly at the same level of mate value as we are. Numerous innovations, however, now allow all of us to sidestep our own genetic limits for the sake of our future offspring.

## SPERM AND EGG DONORS

Since the development of cryogenic freezing of eggs and sperm and in vitro fertilization, which was first successfully performed in 1978, it is now possible to skip the whole dinner-and-date preamble and get direct access to the genes of a desirable partner through sperm and, less commonly, egg donors. An entire industry has sprung up to provide future parents with the access to the genes of a partner whom they will never meet, and in the United States alone there are more than 100 sperm banks to choose from.[72] However, you don't just purchase some random vial of "Joe Sperm"—rather, each individual sample gets a remarkably in-depth marketing treatment. For example, consider the boutique marketing pitches of the largest reproductive tissue service in the United States, California Cryobank, which proudly advertises that "less than 1% of applicants qualify as California Cryobank donors."[73] Each of their vials of sperm— which at the time of writing goes for $740—comes with 24 pages of exceedingly detailed biographical information about the donor that goes far beyond that what is available on any dating website. For example, here's how California Cryobank describes Donor 14289. He is referred to as a "super scientist" who resembles Christian Slater, is an astrophysicist and devoted college instructor, likes to shoot Nerf guns with his brothers, enjoys watching random cat videos on YouTube, grows ultraspicy peppers competitively, buys burgers for homeless people, and enjoys whitewater kayaking, sky-diving, and swimming with sharks. The Cryobank staff further notes that this donor always responds to requests quickly, never shies away from asking questions, and dresses casually with wire-rimmed glasses and button-down shirts. Potential customers can also hear a tape-recorded conversation from the donor, see pictures of him as a child, and, if a pregnancy is successful, can get a keepsake package from him to share with their child.

It's curious why so much rich and detailed information is provided about sperm donors, given that most companies, including California Cryobank, have regulations to ensure that you will never have an actual relationship of any kind with this person. The extravagant marketing of donors lays bare the underlying essentialist rationale of this industry: the implicit message behind these 24 pages of rich description is that all of your donor's characteristics are ultimately to be found curled up in the DNA in the tiny frozen vials that are for sale. Any child conceived with Donor 14289's sperm might also be expected to be kind to homeless people, be good at science, enjoy high-adrenaline activities, prefer a business-casual style of dress, and have an attraction to spicy peppers.

The idea of selling sperm from especially accomplished donors started in California in 1980, when "The Repository for Germinal Choice" was launched. The media quickly dubbed the Repository "The Nobel Prize Sperm Bank," because of the explicitly eugenic goals of its founder, Robert K. Graham, who wished to "stimulate [man's] ascent toward a new level of being."[74] Although in the end, the Repository produced no babies from Nobel Prize winners (as the samples from its three Nobel Prize–winning donors were all apparently sterile, and it turned out that prospective parents weren't all that interested in them as donors anyway), the Repository did advertise that its purpose was to only provide sperm from the most accomplished and genetically fit donors. As was noted in a gripping investigation into the Repository by David Plotz, titled *The Genius Factory*,[75] the most popular donor from the bank, "Donor Fuchsia," was nothing like the stereotypical Nobel Prize winner. Rather, he was well-rounded, "not weird or nerdy," had a high IQ, and was an Olympic gold medalist and an author. Donor Fuchsia made all of his sperm donations in the '80s, but in trying to imagine what a 21st-century equivalent of him might be like, I think of the identical twins Cameron and Tyler Winklevoss. The Winklevoss twins, who purportedly had the original idea for Facebook, are tall, handsome, multimillionaire, Harvard-educated Olympic athletes. If the Winklevoss twins ever became sperm donors, they surely could command top dollar because research shows that they are pretty much exactly what women are looking for in a donor. When leafing through the lengthy catalog descriptions for sperm donors, women place more weight on their donor's health (such as his family's health history), his attractiveness and height, and his various abilities (such as his intelligence or musical talent) compared with women searching for a romantic partner. In contrast, when women are choosing a romantic partner, they tend to place a lot more weight on a potential mate's character (such as kindness, honesty, and considerateness).[76]

This desire to get the best genes is also seen when people are choosing egg donors. Although the American Society for Reproductive Medicine offers guidelines that egg donors should not be paid more than $10,000 (for concerns of undue inducement) for the time-consuming, somewhat unpleasant, and potentially risky procedure of harvesting eggs,[77] this recommendation is frequently ignored. Advertisements have been placed in newspapers of top universities seeking quality donors, and offering as much as $100,000 in compensation.[78] For example, an ad that ran in the *Harvard Crimson* in 2006 sought an attractive, athletic woman, less than 29 years old, with a GPA over 3.5 and a SAT score over 1,400.[79] On average, every 100 points in the average SAT score of incoming university freshmen is associated with an extra $2,350 in the price offered to egg donors.[80]

This targeting of donors with highly desirable characteristics probably doesn't strike you as all that surprising—if you were seeking an anonymous donor, what kinds of characteristics would you be looking for? But the donor market reveals that the quest to find good genes for your children becomes far more straightforward once you move from the inherently challenging task of finding the right life partner, to the relatively simple goal of finding the right genetic donor. The donor transaction circumvents the matching hypothesis because it ensures that there are enough 10s to go around. And while on the surface this seems beneficial, it reveals one potential societal cost of donor banks. The matching hypothesis ensures a highly heterogeneous gene pool across the population—when people are allowed instead to select the traits in donor sperm and eggs, it could contribute to a homogenization of the gene pool by buyers focusing on donors with a rather limited set of desirable traits. Moreover, if it became the norm for everyone to have kids from accomplished sperm and egg donors, then those who opted to stick with having children naturally with their only "average" partners might be discriminated against. They might be seen to have failed to live up to their obligation to produce the "right" kind of children.

### CLONING

It may be a truism that people want to provide good genes for their children, but what if you are convinced that there really aren't any genes better than the ones you already have? If this sounds like you, then you might be attracted to the option of cloning yourself.

Cloning is a relatively straightforward technology, in principle, if not in practice, that involves removing the nucleus from the cell of an individual organism and swapping it with the nucleus of an egg cell. This egg cell, which contains all of the organism's chromosomal DNA, can then be inserted into a womb where it can be "reprogrammed" to act like any other fertilized egg, and develop, at least in theory, to be identical to the original organism.

The idea of cloning became famous in 1996 when Dr. Ian Wilmut and his colleagues at the Roslin Institute in Scotland cloned Dolly the sheep, the first time that a mammal was ever successfully cloned. Never before has a sheep enjoyed as much worldwide fascination as Dolly did. It wasn't a huge leap for people to imagine that what was possible with the sheep should also be possible with the shepherd, and the resulting media hysteria about human cloning revealed the kinds of untenable intuitions that people have about the idea. Newspapers and magazines imagined a future where perhaps people could clone themselves to have a complete set of spare organs so they could continually be updated with brand new parts like a collector's car; children could be replaced if they were lost through tragedy; you could make a double to protect yourself from assassination attempts; an unbeatable basketball team of Michael Jordans could be assembled; people could make copies of themselves so that they could have someone else to do their tax returns for them; men could just be disposed of entirely, as they would no longer be necessary; or, as the organization Christians for Cloning of Jesus proposed, we could clone Jesus so that we would no longer have to wait for his return.[81] Hollywood movies have also taken the idea of cloning to its most absurd ends: for example, the 1978 movie *The Boys from*

*Brazil* described the imagined efforts of Dr. Josef Mengele to clone Adolf Hitler to create an army of 94 Hitlers, identical to the original, to renew the Nazi quest for world domination. The 2000 film *The 6th Day* portrays cloning as a means for gaining a kind of immortality, where people never really die because an individual's cells can be cloned from his or her corpse, and the new body and mind can immediately pick up where the individual had last left off. The 1996 film *Multiplicity* saw cloning as a way to make you more productive at work and home by creating a small army of your selves to divide up the workload.

Where do all these crazy ideas about cloning come from? They certainly don't reflect the science of what cloning would actually entail, as we'll discuss below. If you think your genes are ultimately the source of your underlying essence, however, then duplicating that essence would very well seem to mean creating an identical copy of you, including your physical and psychological characteristics, your memories, and your soul. Such a prospect would indeed pose significant ethical and existential threats to our understanding of who we are. And surely it's because people's understanding of cloning often includes such overwrought and fantastical imaginings, likely fanned by the media and Hollywood's absurd treatment of the topic, that people remain so fearful of it. Indeed, several surveys have found that people are more disturbed by the idea of human cloning than they are by any other prospective genetic technology. These surveys find overwhelming agreement to the statement "cloning threatens the natural order of things."[82] No other technology is as unsettling or undermines our idea of who we are as much as human cloning.

We should contrast these ridiculous ideas about human cloning with what cloning would actually be like. The closest approximation to a clone is an identical twin, who literally is a genetic clone of his or her twin, and not only shares the same genes, but also shares the same experiences in the womb and family upbringing. Although many identical twins certainly do share much in common—case in

point, Cameron and Tyler Winklevoss—they each possess many differences from each other as well, and they certainly don't share a
common mind or soul. Manufactured clones would actually be far
more different from each other than identical twins because they
wouldn't share the same womb or upbringing—they would have different sets of experiences, perhaps separated by years or even generations—which would affect how their genes were expressed, and
their memories and concerns would be the product of their distinct
experiences. But because people think of genes as essences, imagining someone with the same genes as yourself feels akin to imagining
someone who shares your very essence—someone who *is* yourself.

There are no insurmountable technical hurdles for scientists seeking to successfully clone a person, just as they did with Dolly the
sheep. In 2014, scientists cloned human embryonic stem cells, but
they didn't allow the cells to develop beyond that.[83] In 2002, the
Raelians, a French cult that preaches that humans descend from
aliens, claimed to have successfully cloned a human, although they
have never publicly produced the child. But if a human ever was
cloned, how similar might we expect the two clones to be? Perhaps
we can get an idea by looking at the first cloned cat. In 2001, cheek
cells were taken from Rainbow, a calico cat, and these were used
to clone a new cat, christened CC (for carbon copy). The cats certainly do have an overall physical resemblance to each other—however, it is far from a perfect match. Whereas Rainbow has patches of
orange fur mixed in with her black back and white belly, CC has no
orange coloring at all. Moreover, whereas Rainbow had always been
quite reserved in temperament, CC, who had been handled a lot
as a kitten, grew up to be more curious and outgoing. The striking
differences between these two cats that share an identical genome
reveal that development is not just the matter of a genes being read
like a fixed script. CC's different-colored coat was the result of a class
of epigenetic processes associated with random patterns of inactivation of the X chromosome, and her different personality is likely due

to the different experiences that she had while being raised.[84] Most likely, if you were ever cloned, your clone would differ from you in even more diverse ways.

## DESIGNER BABIES

Beyond everything we've discussed so far, our dystopian imaginations, fueled by the richly textured examples described in *Gattaca* or *Brave New World*, perhaps get most carried away when we consider how genetic engineering technologies may be applied to customize a designer baby. Imagine, if you will, being able to specify in advance the set of traits you would like to see in the embryo of your future child. Aside from the obvious preference of ensuring that your embryo doesn't possess any harmful genetic diseases, what if you could dictate your child's personality, intelligence, appearance, or athletic ability? What if having a baby was akin to ordering a computer online, where you're able to specify a seemingly endless series of options, such as which graphics cards, monitor, memory, keyboard, mouse, CPU, and various software packages were installed? Some now worry that this is the future that awaits us, with the recent discovery of CRISPR/Cas9, our most powerful genetic engineering tool to date.

CRISPR/Cas9's humble origins highlight the inherent value of basic scientific research. In 1987, researchers studying some rather obscure questions of little apparent practical value about bacterial genomes first identified some unusual repeating sequences in the DNA of some bacteria.[85] At first, the function of these "clustered regularly interspaced short palindromic repeats" (CRISPR) remained unknown, but further study showed that they functioned as a kind of immune system for bacteria. When a bacterium was attacked by an invading virus, CRISPR allowed the bacterium to create a string of DNA that matched the string introduced by the virus, and this sequence was inherited by subsequent generations. If the same virus ever attacked any bacteria again, then the "CRISPR associated protein #9" (Cas9) would be able to edit the DNA of the bacteria such that it completely disabled the viral DNA. A number of subsequent discoveries led researchers to realize that this same Cas9 protein could function like a search command in a word processor; it could search for a specific line of genomic text, flag it, and then allow researchers to swap the segment of DNA out with another line of genomic text. This needn't be limited to bacterial DNA but theoretically could be applied to any cell in any organism, including those of human embryos. Hence, CRISPR/Cas9 can allow researchers, in principle, to replace almost any line of genetic text with another. Needless to say, this has a vast array of potential applications.[86]

It is well-suited for developing GMO agricultural products because it has the psychological advantage of not requiring any of those uncomfortable transgenic transformations that lead people to imagine fish-shaped tomatoes. Genes wouldn't need to be introduced from other species; instead, the original genome could be edited like a passage of text. But it is CRISPR's potential use in engineering humans that has captured the most attention. Despite a call by geneticists for a moratorium on using CRISPR to alter the human genome,[87] in 2015, a team of Chinese researchers were the first in the world to edit the genomes from some human embryos.[88] The embryos were not viable and never left the test tube, but still the very idea of engineering human embryos was bothersome enough that the article describing this research was originally rejected for publication, purportedly because of ethical concerns.[89] In the same year, British researchers also asked permission from the UK Human Fertilisation and Embryology Authority to edit human genomes,[90] and other researchers demonstrated that it was possible to do multiple genetic swaps to the same genome: a total of 62 genetic swaps were performed on a pig cell toward the ultimate goal of engineering pig hearts to be a better immunological match for transplants in people.[91] It's conceivable at some point that a single embryo could go through dozens if not hundreds of rounds of customized editing across the whole genome, greatly increasing the options for genetic engineering. Although the ethics of editing human embryos continue to be debated,[92] and the precision of the technology continues to be improved, many believe that it's only a matter of time before the first genetically engineered baby is born. As the codiscoverer of DNA James Watson put it: "Once you have a way in which you can improve our children, no one can stop it. It would be stupid not to use it because someone else will. Those parents who enhance their children, then their children are going to be the ones who dominate the world."[93] Likewise, others have speculated (perhaps a little too breathlessly) that this could lead to a new kind of genetic engineering arms race between countries. The fear is that governments will conclude that they'll suffer a competitive disadvantage if other countries are able to outcompete them in improving the genetic potential of their citizens.[94] However, such a future seems unlikely, given people's staunch psychological resistance to the idea. Surveys find that only about 20 percent of American adults are open to the idea of genetic engineering being used to select for desirable traits, and these numbers have not changed over the past decade—the majority of people find the idea to be deeply repugnant.[95]

What kinds of psychological implications would follow from a world with designer babies? Genetic engineering may well represent a kind of back-door eugenics[96]—however, this new kind of eugenics brings with it a different set of concerns compared with the old. One of the most disturbing aspects of 20th-century eugenics was the coercion that it fundamentally entailed, as the state had the final say about who was allowed to reproduce. In contrast, 21st-century eugenics introduces an entirely different concern: the sheer amount of choice that it offers. If you could specify traits for your children a la carte, how would you ever know when to stop? Should you opt for the less anxious temperament? Would your child's life be better if she had blue eyes? Will your child fail to make the basketball team if you don't opt for some fast-twitch muscles? I suspect that prospective parents considering such a decision might go through something like what my wife and I did when we renovated our house—our original plan to replace the carpet in one floor was dropped when a contractor somehow convinced us to reduce the entire ground level of the house to its studs, knocking out walls, building new doorways, and replacing everything along the way. This ultimately provided us with a lovely living space from which we could sit and contemplate how we were ever going to repay the massive second mortgage that we had just taken out. But once we had signed on to the idea of renovating our house, the incremental cost of each additional improvement seemed like a smaller and smaller amount in comparison to the overall budget, and it became easier and easier to say yes to each new addition. How could we not spend the few hundred extra dollars to get the instant hot water dispenser? If people can sometimes lose control in the face of all the options for improving their houses, how are they ever going to show restraint in the face of options to improve their children's lives?

There is a personality trait that would seem to be particularly vulnerable to the dilemma provided by designer babies. Psychologists refer to some people as "maximizers" because they continually seek to get the best possible deal that they can. For maximizers, "good" is never good enough; they keep searching until they can find the best possible outcome. Maximizers are always out to achieve the highest possible standard; they channel surf more in the hope of finding the best possible show that is currently playing, date a longer list of people in the hope of finding Mr. or Ms. Right, and can be left paralyzed by a long menu in a restaurant. Maximizers stand in contrast to "satisficers," who are only interested in finding something that passes their threshold of acceptability; satisficers tend to quit their searches quite early, as soon as they find something that seems good enough. They spend little time wondering if there might have been better possible options that they could have had instead.[97]

The advantage for maximizers is that they are more likely to end up with better-quality outcomes to their decisions. You probably have a maximizer friend who has somehow found the most amazing apartment, got an incredible deal on a used car, or can take you to the best taco joint in the city. But despite the good outcomes that can come to maximizers, they entail a substantial psychological cost: the trouble with pursuing perfection is that you're never able to achieve it. Ironically, despite that maximizers make decisions with objectively better outcomes than satisficers, they are consistently less satisfied with their decisions. They also tend to be less satisfied with their lives and are more depressed. In particular, maximizers are more likely to feel regret after their decisions—they're left with the nagging sensation that, despite their efforts, there probably was an option that was even better than the one they selected.[98]

The notion of having options to genetically engineer your children would seem like a recipe for psychological disaster for those with a maximizing personality. If the quest of a maximizer is perfection, and you have at your hands a potentially infinite number of options to genetically sculpt your future children, how will you be able to avoid the inevitable feelings of regret as you watch your child develop? What if your child isn't a genius, or suffers from some kind of mental illness, or doesn't look like the supermodel that you had imagined? When you look into your child's eyes, and past his glasses, how do you avoid the regret that you should have paid more to ensure that he didn't have any "myopia genes"? Or what if your efforts to boost his height had the unintended side effect of making him more at risk for developing bone cancer? People frequently experience buyer's remorse after they complete a large purchase, questioning whether they really made the right decision. It's hard to imagine a situation where you'd feel more remorse than whether you made the best decisions for your child's genes; and unlike other purchases, if you're not satisfied with this one, you can't return it. Conversely, imagine the reactions of your engineered children. Children already have no difficulties coming up with reasons to resent their parents, but what if they could also resent their parents for the genetic engineering decisions that they made for them? "Why did you give me curly hair?" "Why didn't you make me as tall as the other kids?" "What if I don't want to become the scientist that you engineered me to be?" And how many children would suffer from feelings that they were unable to live up to the high expectations that their parents so clearly held for them when they handpicked their genes for success?

We might also be worried about the kinds of traits that we would ultimately select for our children. Would we all eventually choose the same traits for our kids and be swamped with genetic homogeneity, creating a world full of Winklevosses? Or would we ultimately make choices to engineer children in particular types, such as an artistic type, an athletic type, a scholarly type, and so on? Perhaps through human genetic engineering, we would re-create what dog breeders have done with canines, where we have a set of distinct "breeds" with identifiable traits. And what might we lose if we start to pursue certain genetic types or strive to avoid other ones? There's a common trope that many artists and geniuses appear to suffer from one mental illness or another, such as Ernest Hemingway's depression, Isaac Newton's autism, Jackson Pollock's bipolar disorder, or John Nash's schizophrenia.[99] Would our efforts to engineer our children leave society without the mental illnesses that have spurred some of humanity's greatest creations? And what if we made efforts to weed out genes with problematic outcomes, such as the MAOA "warrior gene"? As we noted in Chapter 7, because this gene is highly pleiotropic and is also associated with a vast list of other traits, many of which are desirable, any efforts to reduce its frequency in the population would likely lead to many kinds of unintended side effects.

However intriguing it is to speculate on the consequences of a world filled with designer babies, I must now confess that this last section has been entirely misleading. Because if there are any traits that we will *not* be able to engineer, it's precisely the kinds of complex traits that we value so much, such as intelligence, attractiveness, athletic abilities, personality traits, and height. All of these traits are massively polygenic, and they are the product of many hundreds, if not thousands, of genes interacting with their environments as individuals develop through their lives. There are no single genes that contribute much at all to any of these traits, so the notion that we could genetically sculpt people to be better on such characteristics appears to be nothing but a pipe dream. Genetic engineering to avoid Mendelian diseases is already with us in the form of PGD, and any future CRISPR/Cas9 engineering of humans would also surely be applied to traits that can be understood through switch-thinking. Perhaps the targeted genes might be broadened to include diseases with unusually strong genetic predictors, albeit less deterministic ones, such as BRCA1's role in breast cancer, or perhaps even APOE's role in Alzheimer's. But because the vast majority of our traits emerge through vastly tangled pathways better understood with web-thinking, the future promised in *Gattaca* remains science fiction.

As mentioned before, we are prone to conceive that all human traits, like intelligence, physical attractiveness, or height, are aspects of our essence, and that they each have their corresponding switch-like genes. This makes it easy to imagine that all of the traits of an individual can be toggled on or off with the appropriate genetic engineering. And this idea that the richness of our humanity could be reduced to something akin to an online shopping experience remains both tantalizing and terrifying. But because most traits don't emerge through any simple switches, most of our imagined futures of genetic engineering will remain just that: imaginary. If most complex traits are best understood through web-thinking, then we need to return to the question posed at the beginning of this journey: How should we be thinking about the role of genes in our lives?

# 9.

# Harnessing Our Essentialist Impulses: How Should We Think about Genes?

**AS THE RESEARCH DESCRIBED IN THIS BOOK** reveals, not all ideas are created equally, or at least not received equally. When we encounter ideas about our genes, we begin to think differently—our essentialist biases take over. When we learn that our genes are involved in our life outcomes, we see the world unfolding in front of us as if it was determined by fate. Thinking about "obesity genes" can render us fatalistic about our weight; considering genetic differences between men and women can make us more sexist; the notion of gay genes can make us more supportive of gay rights, yet more likely to see gay men and lesbians as fundamentally different; ruminating about our own genetic health risks can keep us up at night; thinking about the ancestral origins of our genes can change our sense of identity; pondering the dispersal of different genes across the world can trigger racist thoughts; the idea of "warrior genes" can make us both more forgiving and more frightened of violent criminals; focusing on the genetic foundation of human traits can lead us through the doorway to eugenics; and contemplating the artificially modified genes in our food can make us feel deeply violated. Genes simply captivate us in ways that other concepts do not, and they render us somewhat helpless under their charms. How can we break the spell that they have cast over us?

## The Irresistible Appeal of Genology

Genetic influences seem to be the unchanging and irrepressible forces at our core that ultimately make us who we are; we believe they hold hidden truths about us. Because of this almost mystical quality, people often turn to genetics as they might turn to astrology. Sydney Brenner, a Nobel Prize–winning molecular biologist, called this aspect of genetics "genology." "That's the science that says, 'If you've got this SNP and that SNP, you are going to become a dark stranger!'"[1]

Comparing the predictions made by direct-to-consumer companies to astrology might seem surprising—isn't the former grounded in science, whereas the latter is simply superstition? And surely the fact that people largely see truth in the genetic predictions that they get from direct-to-consumer companies is proof of their vastly superior accuracy, right?

Not at all. It turns out we are psychologically equipped to unquestioningly believe in the predictive power of genetic feedback, just as many of us remain convinced by astrology, even though that feedback is sometimes deeply inaccurate. For example, one reason we so often see truth in both horoscopes and in genetic feedback is because of a psychological bias known as the "Barnum effect," after the circus hoaxer P. T. Barnum. You can see this effect in the following personality test. Answer whether you agree or disagree with the ten items below, and then tally up how many of those items you agreed with.[2]

1. You prefer a certain amount of change and variety and become dissatisfied when hemmed in by restrictions and limitations.

2. At times you are extroverted, affable, sociable, while at other times you are introverted, wary, reserved.

3. You have a tendency to be critical of yourself.

4. Disciplined and self-controlled outside, you tend to be worrisome and insecure inside.

5. At times you have serious doubts as to whether you have made the right decision or done the right thing.

6. You have a great need for other people to like and admire you.

7. You pride yourself as an independent thinker and do not accept others' statements without satisfactory proof.

8. Some of your aspirations tend to be pretty unrealistic.

9. Security is one of your major goals in life.

10. You have a great deal of unused capacity which you have not turned to your advantage.

OK, so how well did this personality test capture your feelings? How many of the 10 items did you agree with? Maybe all of them? Perhaps these items seemed to reveal your innermost self. Or perhaps you saw the trick straightaway: these items seem so characteristic because they are so broad that they are true of virtually everyone. It's likely not difficult for you to think of some specific examples from your life that fit with each of these general and commonly experienced descriptions. This is how the Barnum effect operates, and you can see it in astrology too; the different zodiac signs rarely offer you concrete and specific information that could be refuted, such as "You always complete your taxes at the very last minute," or "You prefer the taste of broccoli over carrots." Rather, the characteristics of zodiac signs are more usually described in the broadest of brush strokes that can seem consistent with virtually everyone. For example, my sign, Virgo, tells me I am particular, logical, practical, reflective, and thoughtful. Perhaps you can see these traits in your own character, even if you happen to be a Pisces, or Aquarius, or any of the other signs.

Remarkably, despite our impression of genetics as a cutting-edge technology, many gene associations are described in ways that are very similar to these horoscopes. For example, one study found that a particular variant of the RGS2 gene (specifically, the CC allele at rs4606), is associated with a more anxious temperament,[3] although the associations are somewhat preliminary at this time. Do you think you might have an anxious temperament? Feelings of anxiety are certainly very common, and so is the CC allele at rs4606—in fact, more than 50 percent of people have this allele. This particular marker is included in the 23andMe test, and it's illuminating to read the discussion thread associated with this marker at the 23andMe website.[4] Almost all the individuals on the thread who have the CC allele point out ways that anxiety is part of their lives, and the vast majority of them seem to view this gene as the cause of their anxiety. Others even say that because their parents are CG they want to get their siblings tested, as they feel they could predict which siblings have which alleles. Likewise, the cognitive scientist Steven Pinker described how he made sense of learning about himself from his own genome. "I soon realized that I was using my knowledge of myself to make sense of the genetic readout, not the other way around. My novelty-seeking gene, for example, has been associated with a cluster of traits that includes impulsivity. But I don't think I'm particularly impulsive, so I interpret the gene as the cause of my openness to experience."[5] When predictions are made about such broad traits, the combination of the Barnum effect and our tendency to view genes as ultimate causes can make "genology" explanations extremely compelling. Often you're going to find a match between your self-perception and some aspect of a trait associated with your genotype. When that happens, you might feel that this genetic oracle has seen right through to your core.

You're also vulnerable to genology because most of the genetic feedback you get is simply beyond your ability to confirm or reject. You certainly don't know the probabilities that you're going to develop any diseases, nor do you know, aside from family lore, where your distant ancestors really came from. Without any proof to the contrary, you have no reason to doubt most of the genetic predictions that you receive.

There have been some prominent genetic testing misfires, however, that reveal just how tenuous the predictive power of genology can be. For example, the director of the National Human Genome Research, Francis Collins, was predicted by 23andMe to have brown eyes, but his are, in fact, blue.[6] Likewise, Steven Pinker notes that his 23andMe results suggest that he has an 80 percent chance of being bald,[7] which is ironic because his famous locks have earned him membership in the "Luxuriant Flowing Hair Club for Scientists."[8] These kinds of falsified predictions can decrease your faith in the accuracy of genology, but between the Barnum effect and the many traits that are not so easily falsifiable, these false predictions will probably seem rare enough to be largely discounted.

So what can we do to resist the misleading appeal of genology and to think about our genes in more rational ways? Below I describe some steps to help disarm the most harmful consequences of our essentialist biases.

## Distinguishing between Strong and Weak Genetic Influence

First, it is critical to keep in mind that not all genetic influences are the same. We must attend closely to the *magnitude* of influence; with genetic effects, size matters. As we've noted, there are many genes involved in a number of rare diseases, such as Huntington's, Harry Eastlack's FOP, or cystic fibrosis, that have close to 100 percent predictive accuracy. In addition, there is a second tier of genes with lesser but still quite strong influences, such as the BRCA1 gene that Angelina Jolie carries, which has a predictive risk of 40 percent to 60 percent for breast cancer. With risk magnitudes of this size, our usual switch-thinking bias actually comes close to reflecting genetic reality.

As we've seen throughout this book, however, we often apply the same switch-thinking logic to cases of weak genetic influence in which the individual genes are each making the most tenuous of contributions. And the vast majority of our traits and conditions are weak cases, emerging from the confluence of many, many genes interacting with our experiences and epigenetic processes. Far more often than not, our switch-thinking leads to grossly incorrect conclusions and unwarranted fears.

Once you've focused on the *magnitude* of genetic influences, the natural next step would be to match the strength of your emotional reaction to the strength of the genetic influence. However, matching your psychological reaction to your genetic risk is not at all an easy thing to accomplish. It's challenging both because probabilities are not an intuitive way of thinking, and because we often think of risk factors that exist within us, in our essences, as an inherently inescapable kind of risk, regardless of their magnitude. It's just frightening, no matter how small my increased risk, to consider that my genes make me more vulnerable to Parkinson's—or even worse, that I have "the Parkinson's gene." It feels like my very essence is conspiring against my health.

Another problem with risk estimates is that serious conditions subjectively *feel* more likely to occur than less serious ones, even if the actual probabilities are held constant.[9] For example, if tomorrow there's a 20 percent chance of rain in your neighborhood, you might conclude that it's not that likely to rain. In contrast, if you learned that tomorrow there was a 20 percent chance of a tornado in your neighborhood, you might see a tornado as a disturbingly likely outcome. The severity of the tornado makes its likelihood seem worse, even though the probability is equivalent to the rain forecast. And because severe conditions tend to be thankfully rare, any chance greater than zero of getting one, such as my 2.1 percent chance of developing Parkinson's, can feel far more likely than its small probability value indicates. This mismatch between our subjective sense of risk and our actual risk causes genetic counselors, in particular, to struggle with communicating genetic risk estimates accurately. Indeed, studies find that people often can't recall what their actual genetic risk level is, even after receiving genetic counseling informing them of their objective probabilities of developing certain diseases.[10] One of the founders of the field of genetic counseling, Sheldon Reed, stated that many clients "don't oftentimes distinguish between one chance in four and one chance in one hundred. It's all the same to them."[11] Both of these kinds of risk probabilities are perceived by some people as "it could happen to me."

This mismatch between people's subjective worry and actual risk probabilities is paralleled by many genetic counselors themselves, who often view *any* level of probability that's greater than 0 percent as a risk worth being concerned about. One survey found that a 2 percent chance of having a pregnancy with a genetic disease was considered a "high risk" or even a "very high risk" by about half of genetic counselors,[12] and many genetic counselors responded that "no risk is tolerable."[13] No matter which side of the table we're on, it seems our actual risk probabilities get drowned out by our fears.

Benjamin Cheung and I investigated how much this insensitivity to the magnitude of genetic risk would affect how people respond to other genetic risk factors, specifically, to the idea of "criminal genes." We had American adults read about a case where a college student named Patrick killed another man with a knife in a fight.[14] One group of participants read that Patrick possessed a gene that was associated with a 25 percent increase in the likelihood of violence. Another group read the exact same vignette—with the exception that the same gene was instead associated with a *400 percent* increase in a likelihood of violence. A third group read the same vignette without any mention of Patrick's genes. The participants were then asked to consider whether an insanity defense was appropriate to reduce Patrick's sentence. The first two groups, who learned about Patrick possessing genes associated with violence, thought the insanity defense was more appropriate than the third group did. Yet the magnitude of the genetic associations had virtually no impact on them. They responded the same whether the genetic risk of violence was only 25 percent or was a whopping 400 percent. All that seemed to matter was that because Patrick had a genetic risk—any risk—he wasn't seen as fully responsible for his crime. When we hear that genes are involved, it's as if we just hear that "it's genetic," and whatever discussion there is about the actual strength of the genetic influence goes largely unheard. We respond as though the genetic switch has been turned on.

What makes these oversize reactions even more problematic is the uncomfortable fact that genetics research is not such a precise enterprise to begin with. Many, if not most, genetic discoveries that are published in the best journals and then reported in newspapers around the world ultimately turn out to be red herrings. An editorial in the *American Journal of Medical Genetics Part B* noted that "It is no secret that our field has published thousands of candidate gene association studies but few replicated findings."[15] Similar problems exist for genome-wide association studies. Many are highly likely to reveal spurious gene associations by mere chance,[16] and even if the identified associations prove to be reliable, they are often too small to have any consequential impact on our lives.[17] The journalist David Dobbs characterizes the arc of genetic research as "alluring hope, celebratory hype, dark disappointment."[18] Unfortunately, given the penchant for both academic journals and newspapers to get excited by new findings, we all hear about the amazing initial discoveries but often never hear about the subsequent failures to replicate or the ultimate realization that the reality is far less interesting. For example, in 2009, a genome-wide association study found a particular SNP at rs4307059 that was identified as a risk factor for autism and was said to have a probability of being identified by chance of less than .00001 percent—suggesting a highly reliable effect.[19] The *New York Daily News* ran an article on this research and noted how this newly identified gene increased one's risk for autism by 20 percent.[20] Nonetheless, a subsequent paper published that same year found that people that possessed this same genetic risk factor were actually somewhat *less* likely than average to get autism, yet this failure to replicate didn't receive any media coverage.[21] Such failures to replicate are extremely common, which means that not only are we overreacting to a lot of genetic influences that are really small in scope, but we are often overreacting to findings that don't even exist. These phantom fears are akin to being afraid of zombies attacking you in your sleep.

Why do initial genetic findings so often fail to stand the test of time? There are two answers to this. First, the field of genetics shares the same problem as the rest of the sciences: scientists are rewarded in their careers for getting their work published, which leads to many "discoveries" that are published in academic journals and covered in media outlets before it can be determined whether those findings are actually reliable. There is a sizable leap between discovering something new and in demonstrating that this finding is real and can be reliably reproduced by others.

A second reason that many genetic findings do not replicate is more specific to genetics research. They don't replicate because, reassuringly, most phenotypes are the product of weak genetic influence, and therefore the vast majority of genetic associations are by themselves extraordinarily weak in magnitude and are thus extremely difficult to reliably observe. They are like a faint whisper amidst the loud din of all of the other influences on a life outcome, and when researchers think they have detected the whisper, often they have only identified something that was actually a phantom.

Consider the so-called "novelty-seeking gene," which is identified by a variable portion of the DRD4 gene. This is one of the most studied genes in behavioral genetics,[22] and has received a great deal of media interest, where it goes by a number of other labels, such as "the liberal gene" or "the risk-seeking gene." In 2015, Land Rover referred to this gene in a commercial for the company's Discovery Sport vehicle, calling it "the adventure gene," and stated in no uncertain terms that "if you have it, you will be more adventurous and take more risks." Given the vast amounts of research and media interest on this gene, it would seem to be a pretty big deal to know which allele of this gene you possess. But how much of your novelty-seeking/adventure/risk-seeking does this extremely well-studied gene actually account for? Approximately 0.1 percent![23] And even this number is likely an overestimate of the gene's influence.[24] Trying to discern someone's novelty-seeking by looking at this one gene is akin to having a single piece of a 1,000-piece jigsaw puzzle and trying to imagine what the complete image would be like.

Anytime you hear about a single genetic variant that predicts a condition, the best thing you can do is focus on the magnitude of the association. If a single variant has a large influence on some condition (which is rarely the case), then this is something important for you to note, especially if you are a carrier of this variant. Also, because this variant would almost certainly be predicting a condition that is a rare genetic disease, you should speak to your physician for guidance. But the vast majority of genetic predictions about you that are provided by direct-to-consumer companies are of such small magnitude, and are hidden deep within such an impenetrable cloud of uncertainty around the precision of those estimates, that you should simply just ignore them. Better yet, you could just save your money and avoid them entirely. These flimsy wisps of genetic influence will do nothing but lead you to worry, even though they have absolutely no bearing on your life.[25]

## The Essence Isn't the Answer

It is critical to remember that our attraction to essences is a bias in the ways that our minds process information. When we focus on essences as the drivers of our lives, we fail to appreciate the huge impact that our environments can have on us. And this bias toward essences can have a significant impact not only on our own life decisions, but also on decisions to fund science.

Consider the case of autism. Autism is frequently called an "epidemic"—as many as 1 in 68 children born in the United States are now diagnosed with the condition.[26] There are huge costs both for the expense of the treatments to improve the lives of those diagnosed, and also in terms of the social costs associated with the difficulties of being autistic. So what is the best way to understand the cause of autism? Well, one approach would be to focus on the genetics underlying autism, and indeed scientists have embraced this with a vengeance.[27] In the first decade of the 21st century, approximately one billion dollars was spent on investigating the genetics of autism. The results of these investigations point to a glaring case of weak genetic influence. There are no single autistic genes, only a vast web of hundreds of variants which interact in unspecified ways to predispose one toward developing autism.[28] In contrast, in that same first decade of the 21st century, only about 40 million dollars was spent on studies of the environmental influences on autism. That's a 25-fold difference between the research funding on genetic versus environmental influences.[29] Can autism be explained 25 times better by genes than by the environment? Will an understanding of the role of genes in autism lead to more effective interventions than those based on the environment? I, for one, remain doubtful. There have been no significant changes in the human genome since autism cases have exploded, so it seems quite unlikely that this new "epidemic" can largely be accounted for by genetics. But the notion of a genetic basis of autism resonates deeply with our essentialist biases.

The funding decisions for autism are by no means unique. For example, in 2011, the National Institutes of Health spent approximately four times as much on genomics research than it did on all of the basic behavioral and social sciences combined.[30] Clearly, there have been many advances in genetics research as a result of this generous funding, and both the academy and society stand to benefit greatly from all of this new knowledge. However, there is a clear cost when the center of gravity for funding moves so dramatically toward one field at the expense of others. It changes the kinds of questions we ask and the interventions we pursue. As the psychologists Jay Joseph and Carl Ratner put it, "Genetic-determinist ideas divert society's attention from these environmental conditions and shift blame onto people's brains and bodies. Even in the case of medical disorders such as Type 2 diabetes, where poverty and malnutrition are well-known causes, supporters of genetic determinism continue to press for research dollars to be directed toward genetic research, as opposed to improving social and health conditions."[31]

Funding research that targets genetic causes for conditions will only provide genetic answers for those problems. And because our lives unfold as our biology (of which our genes are only a part) interacts with a set of experiences across a developmental timeline, those genetic answers can at best provide only part of a potential solution. And while there is certainly much we can learn from genetic research, it doesn't even necessarily follow that genetic answers are the most useful kinds of answers. Genetics research is often described as the holy grail, with a rich bounty of medical benefits that lie in the offing. That might be the case. But it's useful to keep in mind that the history of genetics research has been made up of constant promises that the next big cure is right around the corner. As the science historian Nathaniel Comfort puts it: "Studying genetics and popularization over the last century or so has led me to the surprising conclusion that genetic oversell is independent of genetic knowledge. We see the same sorts of articles in 2014 as we saw in 1914. Neither gene mapping nor cloning nor high-throughput sequencing; neither cytogenetics nor pleiotropy nor DNA modification; neither the eugenics movement nor the XYY controversy nor the debacles of early gene therapy—in short, neither methods, nor concepts, nor social lessons—seem to make much of a dent in our preference for simplistic explanations and easy solutions."[32] The appeal of essentialist answers to all life's questions has led to much overpromising of results from genetic research since its earliest origins.

The history of delivering on genetic hype and promises is far less encouraging. A good case in point is cystic fibrosis. The disease has a straightforward cause—mutations in the CFTR gene, although there are more than 1,000 different kinds of pathogenic mutations along that gene. This cause was identified in 1989, yet more than a quarter of a century later, we're still struggling to improve the treatment of cystic fibrosis on the basis of this genetic cause,[33] although there are a number of treatments that are in development that target particular mutations along this gene, such as the drugs ivacaftor and lumacaftor.[34] As one of the original researchers who identified the CFTR gene, Jack Riordan, put it: "The disease has contributed much more to science than science has contributed to the disease."[35] Contrast the relatively straightforward monogenic cause underlying cystic fibrosis with the hundreds of different genetic variants that have been identified with autism or schizophrenia. If we struggle with gene-based treatments for cystic fibrosis, what kinds of genetic interventions will emerge from our understanding of the tangled web of genetic influences involved in these conditions?

I don't mean to be overly pessimistic here. There are numerous medical interventions that have emerged from our rapidly growing understanding of our genome. In certain cases, we can now make actionable life changes as the result of a strong genetic risk, such as Angelina Jolie's prophylactic mastectomy and oophorectomy after learning that she was a carrier of a pathogenic variant of the BRCA1 gene. And although the bounty of gene-specific pharmaceuticals has been much slower to emerge than the hype from the Human Genome Project suggested,[36] many such drugs have emerged, in particular, for targeting various cancers. For example, vemurafenib has shown much promise for treating some kinds of melanoma,[37] and imatinib is effective for treating some kinds of leukemia.[38] But scientific funding is very much a zero-sum game, and by continuing to devote so much of our available resources into genetics research, we're diverting funds away from efforts to reveal the very important ways that our experiences and upbringing can also lead to harmful health outcomes. For example, social science research has found that lifespans among the elderly can be improved by providing them with more control over decisions in their lives,[39] coronary heart disease can be effectively reduced by specific lifestyle interventions,[40] Type 2 diabetes is more effectively treated by lifestyle interventions than it is by the leading diabetes medication,[41] guided exercises in writing about traumatic life events can improve lung function among asthmatics and can reduce disease severity for those suffering from rheumatoid arthritis,[42] mindfulness meditation can improve symptoms among HIV-infected adults,[43] and group therapy can increase survival rates of breast cancer.[44] There are multiple ways to treat medical conditions, and although gene-based treatments may resonate with our essentialist biases, they are not the royal road to medical miracles. Our lopsided focus on gene-based treatments over other options, especially if those treatments don't live up to the hype, comes with significant costs.

## How to Talk About Genes

Our fatalistic ideas about genes certainly aren't mitigated by the ways that we typically talk about them. The language we use when discussing genetics shapes our thinking about the influence of our genes. So a key first step in reducing our essentialist biases is to examine the problematic ways that we talk about genes.

*Avoid Labels for Genes.* Our genes are part of a complex interactive biological system that code for particular proteins, and express those proteins in response to any signals that they receive from our experiences or other biological events. That's it—that is all they do. Describing genes in these rather dry and prosaic terms most likely doesn't elicit much of an essentialist reaction from you, and that's a good thing. Unfortunately, this isn't how we usually talk about genes, especially when they're discussed in the media. Rather, we have a penchant for trying to represent a gene's function by giving it a pithy little label, such as calling it "the breast cancer gene." These labels serve as a useful shortcut to reduce the inherent complexity of genetic mechanisms to something we can understand: the gene causes breast cancer. But the labels also seem to imbue genes with a sense of purpose, and often a rather nefarious-sounding purpose at that.

Take the DRD4 gene that we were discussing earlier when we talked about novelty-seeking. This gene has been branded with many different labels, and one label in particular has received much media attention: "the infidelity gene." Some of the headlines that have covered this research include "Too many one-night stands? Blame your genes." Or "The Love-Cheat Gene: One in Four Born to Be Unfaithful, Claim Scientists."[45] With this label, the gene no longer sounds as innocuous as just a string of nucleotides that codes for proteins—it now seems to not only have a clear objective, but even perhaps the power to override your conscious will when it comes to cheating on your partner. Since the media coverage of this gene, I wonder how many uncomfortable conversations have begun with "Sorry babe, you know I have the infidelity gene."

Genetic labels have been mindlessly applied to all kinds of genetic discoveries, with little consideration for how they affect our thinking. If you type the phrase "scientists find gene for" into Google, you get a list of over 36,000 hits, including the "genes for" cocaine addiction, divorce, sweet tooth, and religiosity.[46] These labels strongly convey the sense that genes are "for something"—they each seem to have a specific purpose. The DRD4 gene is for causing infidelity, and the APOE gene is for causing Alzheimer's. But, really, when you get down to it, genes aren't for anything at all. Even the HTT gene that Woody Guthrie possessed, with its very strong association with Huntington's disease, is not for causing Huntington's—it's for expressing a protein that is involved with the transport of materials within cells. Huntington's disease emerges when the biological system that includes variants of the HTT gene malfunctions. Calling the HTT gene the Huntington's gene is like answering the question, "What are prostates for?" with "They're for getting prostate cancer."

The sociologist Peter Conrad criticizes the media for discussing genetics research with what he calls an OGOD framework: "One Gene, One Disease."[47] It's not surprising that this is the way genetic influences are so often discussed. We are cognitive misers—we do not engage in excess mental effort unless necessary—and so we prefer simple stories. What could be more simple than a one-to-one correspondence between the challenges in your life and a set of matching genetic switches? If you're novelty-seeking, then you must have the novelty-seeking gene, and if you suffer from depression, then you must have the depression gene. And so on. But, of course, this OGOD framework is completely wrong for making sense of our lives. It's hard enough to gain a sense of understanding of the vastly complex and interactive biological systems in which our genes operate without these genetic labels triggering our switch-thinking tendencies. Both scientists and the media need to resist the urge to label genes in ways that misrepresent how they are involved with phenotypes.

*Common Gene Metaphors Give Genes Undue Power.* Another key way that our understanding of genes is shaped by our language is through the metaphors that we use to explain genes. We are heavily dependent on metaphors to understand ideas, and we especially rely on them for challenging ones. Metaphors aid our understanding by allowing us to swap out something we don't necessarily understand with something we do; we seek to find out ways that something that might be unfamiliar to us (such as, say, learning for the first time about the American presidential primaries) has similarities to something that was already familiar to us (such as a boxing match, a marathon, or an episode of *American Idol*).[48] But because we rely on these metaphors to make sense of the unfamiliar, our choice of metaphor can have an enormous impact on how we think about things. If we think of primaries as like a boxing match, then it brings to mind how combative they can be. If we think of them as a marathon, it emphasizes just how long and drawn out they are. Or if we think of them as akin to *American Idol*, it can make them feel more like a superficial popularity contest. Each metaphor brings to mind different aspects of the primary process, and each makes us understand it in very different ways.

Our understanding of genes is similarly bound up by the metaphors we choose. And this represents a special problem: not only do none of the widespread metaphors for genes reflect an accurate description of how they work, but they also all put us into a fatalistic mindset. Consider the most common metaphors that are used to describe genes: they are "blueprints," "recipes," "computer programs," or "the holy grail."[49] Each of these conveys a slightly different understanding, and each is misleading in its own way. If genes are like blueprints, then they must hold all of the information to realize the final plan. If you are in possession of a complete set of blueprints, then you should be able to re-create, down to the smallest detail, the exact construction as envisioned by an architect—perhaps God, in the case of our genes. If genes are recipes, this suggests that all the information necessary to properly create your meal is there from the outset. There will be no more instructions to encounter along the way, and as long as the cook is competent, you can expect the same creation each time. If genes are like a computer program, this conveys that they are

fundamentally units of information, and the program will be executed in an identical way each time it is run. And if genes are akin
to the holy grail, a full genome sequence might suggest that you have
established direct communication with God—you now possess all
the answers to all the questions about you.

We can relate to all of these metaphors, even though they each
provide a different understanding of how genotypes ultimately lead
to phenotypes. However, all of these metaphors share the same shortcoming: they each render your genes as the commander in chief.
None of these metaphors conveys that genes are just one part of a
complex and interactive biological-environmental-ecological system.
Genes play a significant role in the creation of proteins, but much of
the information that guides the creation of these proteins comes from
outside the DNA, and much of this information is encountered as
the organism develops, rather than being there from the beginning.
We can see the limitations of these metaphors in our discussions of
the differences between Rainbow the cat and her cloned daughter
(Chapter 8), or how the average Dutch man of the 21st century is
6 inches taller than his 19th-century ancestors (Chapter 2), or how
a person with an identical twin who has schizophrenia has only a
48 percent chance of developing schizophrenia also (Chapter 4). In
making sense of each of these examples, we need to consider information that lies outside of the genomes.

I suggest another metaphor for genes should be added to the mix
to emphasize their interactive nature. Your genes are like dance partners. By themselves, they can't do much of anything—they are just
standing in the corner, looking awkward. They are entirely dependent on several other things for them to be able to function as a
dance partner. First, they need to hear music, which is akin to the
experiences that we encounter from our environment. The music
guides the moves of the dancer. And their responses to that music
are going to be shaped by their preferences for different genres, and
memories for particular songs, which they have acquired over their

lifetime—much like the way our developmental history shapes our brains to influence how we interpret events in our environments. And, of course, our genes also need a dance partner—someone to both lead them and for them to respond to, and this dance partner represents the rest of the intracellular and extracellular mechanisms that make up our biological system. When all of these are present, we see a beautiful dance. The genes are a necessary part of the dance, but so are all the other components, including the music, our musical preferences, and our dance partner. None of these can produce a dance in isolation, and it's only when they're all present that the dance emerges in its full glory. Thinking of genes as dance partners forces us to attend to how they respond, not just how they lead.

Of course, just like the metaphors of blueprints, recipes, computer programs, and the holy grail, this metaphor of genes as dance partners doesn't do justice to all aspects of genetics. Nonetheless, at least a dance partner metaphor doesn't lead to fatalistic thinking about genes.

## Embrace the Complexity

Changing the way we talk about genes can help to reduce our essentialist biases. But clearly this is not enough—we still embrace switch-thinking about genes even in situations when less problematic language is used. I think our penchant for switch-thinking explanations dates right back to Mendel, whose first discoveries with peas revealed direct one-to-one correspondence between single genes and corresponding phenotypes. One change in a gene, and you'd get a wrinkly pea; another change and you'd have a yellow color. This is how students first learn about genetics in high school, and it certainly seems like a sensible place to start. Of course the problem is that it's not only an entirely inaccurate accounting of how the vast majority of genotype–phenotype relations emerge, but it also leads people to view the world as made up of essences, with all of the problematic baggage that frequently comes along with that view.[50]

That said, biology is vastly complex, and perhaps the best way to teach people about complex scientific ideas is to start them off with a simplified primer before adding other layers of complexity. Other sciences are sometimes introduced to students in highly simplified ways that bear the cost of misrepresenting their accuracy. For example, when students first learn about atomic physics, they learn that the atom is akin to our solar system, with a solid sun-like nucleus made up of protons and neutrons, being orbited by planet-like electrons. It's a message that is far more easily understood than the probabilistic and inherently uncertain nature of a quantum understanding of the atom that students learn in later courses. So maybe communicating genetics should be akin to communicating physics, where inaccurate but comprehensible explanations are offered as "good enough" placeholders until students have the necessary background to understand the far more complex but accurate description of the nature of reality.

But there's a key difference between the complexities of the atom and the complexities of our genomes: we don't make life decisions on the basis of our understanding about how subatomic particles operate. Incorrectly believing that the atom is akin to a miniature solar system has no consequences in our daily lives. In stark contrast, incorrectly believing that genes are like switches leads us to become fatalistic, and can result in increased sexism, racism, and irrational fears about our future disease risks. When it comes to our genomes, I submit that it's far better to highlight the difficulty in understanding the complex machinery of genetics than it is to give people a false sense of understanding that leads them to make potentially costly decisions in their lives.

Can people's essentialist responses be reduced if genetic effects are described in their full complexity rather than in simple switch-thinking ways? Although studies that describe the intricate richness of how genetic influences interact with experiences have thus far yielded mixed evidence for being able to reduce people's essentialist biases,[51] another avenue of research points to a way out of our fatalism. My students and I contrasted the essentialist responses in various studies between people who had received a lot of education on genetics (such as biochemistry majors) and those who had very little such education (university students in other fields, or American adults who signed up to participate in online research). The results have been clear: the more knowledge people have about genetics, the less essentialist they are.[52] People with greater understanding about genetics are less likely to agree that genes determine life outcomes, less likely to subscribe to eugenic beliefs, and are in general less racist and less sexist. This is encouraging news, as it suggests a clear route for avoiding the harmful consequences of switch-thinking: just start to learn more about genes! You've just read this book, so that's a start. But don't stop here—I encourage you to go and read other books about genetics, or even to take any free online university courses about genetics. And keep a closer, more skeptical eye on news stories that you encounter about genetic findings.

## Know When to Fight—and When to Embrace—Your Biases

I recognize that saying that more genetics education is the answer to our essentialist problems is likely too idealistic. People are busy, and not everyone has the time or motivation to learn more about genes. Are there any easier solutions?

Certainly—and this is where the value of learning about psychology, and more particularly, of learning about genetic essentialism kicks in. The field of social psychology has largely been about documenting the irrational and harmful ways that people tend to think. The research finds, for example, that people tend to be far more prejudiced than they would like to be; that people can be callous and unhelpful to others when they encounter strangers with problems; people often make poor decisions for themselves and others; and people can choke under pressure when they become aware of other people's negative expectations for them. All in all, social psychology paints a rather pessimistic picture of the limitations of people's typical ways of thinking.

However, the encouraging news is that many of these faulty ways of thinking can be corrected.[53] We are not limited by the fact that our minds are prone to view the world in biased ways. We can overcome many of these biases with training.[54] And indeed, much research finds that overcoming our biases need not be a lot of work—we just need to be aware that our thoughts might not always be rational. Just knowing about the existence of the kinds of biases that we are all susceptible to can be half the battle.[55] So by learning about genetic essentialism, you should be better protected from your own biases.

And, as I write this book, I too see how my own ways of thinking about genetics have been shaped by learning about essentialism. The truth is, whenever I encounter information about my genes, my first gut-level response is as essentialist as anyone else's. When I perused my 23andMe feedback, I had mild anxiety attacks when I learned of my heightened risks for conditions like prostate cancer and Parkinson's; I was stunned for a while to learn from 23andMe that my maternal ancestry might come from China; and learning of my lack of IQ genes gave me a brief run of self-doubt. But by realizing that these switch-thinking reactions are irrational, that they indicate my deep-set preference to explain the world in terms of underlying and invisible essences, I'm able to discount these reactions, and focus on the questions with a more rational mindset. By being aware that I have these biases, I can disarm them. They no longer control me.

But of course, essentialist reactions are not always harmful; I'm optimistic, for example, about the ways that learning about "gay genes" makes people more accepting of gay rights, and how learning about "depression genes" can make people less likely to blame their depressed friends. This suggests an intriguing approach to managing our switch-thinking biases: perhaps we can, through conscious reflection, *choose* to disarm only the essentialist thoughts we have which are problematic. I feel no need to correct my desirable essentialist reactions, and, in fact, I'm quite happy to embrace them. For example, my 23andMe feedback informs me that I have no working copies of the alpha-actinin-3 in the ACTN3 gene. This supposedly makes it unlikely that I'll ever be a world-class sprinter, which somehow didn't surprise me. On the other hand, 23andMe informs me that this variant of the ACTN3 gene actually makes me more likely to become a world-class endurance runner. And because I do run the occasional marathon, I've learned to put this genetic knowledge to good use. When I get to the point in the race when I've passed the 20-mile marker, and I feel nothing but a wall of pain and hopeless despair that I can't run another step, I think about my 23andMe feedback and I say to myself "I have the endurance gene. It's my genetic destiny to finish this race." And sure enough, those thoughts help to power me through to the finish line. Essentialist thinking is a powerful force indeed. In this new genomic age, wield it carefully.

# Acknowledgments

As with all big life tasks, this one emerged through a lot of collaboration, and I'm indebted to the countless people who helped me with this book. First, I thank my grad students who helped me with the many studies that provided the backbone for this book, most notably Ilan Dar-Nimrod and Benjamin Cheung, as well as Wren Gould, Damian Murray, Cermet Ream, Matthew Ruby, and Anita Schmalor. Other students in the Culture and Self Lab were also instrumental along the way, both in talking with me about the ideas, and in providing feedback on chapters, with particular call-outs to Michael Muthukrishna and Dan Randles. I'm also grateful to John Monterosso, who, as a graduate student at the University of Pennsylvania, first planted the seed of the idea of genetic essentialism in my head more than 15 years ago. A few lab managers also helped me hugely by tracking down an endless series of books, chapters, and newspaper articles—thanks to Adam Baimel, Colin Xu, and Brittany Hathaway. And Kokoro Research Centre in Kyoto provided the most inspiring environment to start working on this book—thanks in particular to Yukiko Uchida there.

I am especially grateful for having such a wonderful set of colleagues at the University of British Columbia, with whom I've had the chance to talk about the ideas in this book, and about the world and human nature more generally. Whether those discussions were on a chair lift at Whistler, at a pub, or at our weekly workshop, I always learned a lot and felt inspired by their company. In particular, my buddies at the Human Evolution, Cognition, and Culture Centre, Mark Collard and Ted Slingerland, always gave me lots to think about, and my countless lunches with Joe Henrich and Ara Norenzayan especially helped in sharpening many of the ideas in this book. And I've had many other stimulating conversations with my other brilliant colleagues in the Psychology Department, including Karl Aquino, Jeremy Biesanz, Sue Birch, Frances Chen, Liz Dunn, Todd Handy, Kristin Laurin, Mark Schaller, Toni Schmader, and Jess Tracy.

My graduate advisor, Darrin Lehman, got me started on being an academic, and he continues to help in so many ways, including by bringing together the collection of great minds at UBC. I wouldn't be here today without him.

The research I report in this book was all funded by the Social Sciences and Humanities Research Council of Canada—I'm grateful for all the support that I've received throughout my career.

This book has benefitted enormously from many colleagues and friends who read and provided feedback on chapters along the way, including Coren Apicella, Michael Coughtrie, Joe Henrich, Matt Lorinz, Dick Nisbett, Wendy Roth, Laura Specker-Sullivan, Vanessa Silva, Eric Turkheimer, and Carles Vilarino-Guell. I also learned much from discussions with Jehannine Austin, David Dobbs, Adam Frankl, Anna Lehman, and Corey Nislow. My original proposal received some sage guidance and writing tips from Jon Haidt and Ethan Watters.

I'm particularly indebted to my agent, Max Brockman, who somehow saw the potential for a book in my proposal. And my editor at Norton, Jeff Shreve, contributed hugely in making this book a more accessible and compelling read—thanks, Jeff, for the many drafts with all of your helpful encouragement along the way.

And of course, thanks to Christine, for supporting and inspiring me through this all. My kids, Seiji and Kokoro, always delight me in letting me see firsthand how genes and experiences can combine so amazingly.

# Notes

## 1. Introduction

1. Transcript from Bill Clinton's speech on June 26, 2000. Retrieved from http://transcripts.cnn.com/TRANSCRIPTS/0006/26/bn.01.html.
2. For a provocative take on future genetic engineering possibilities, see Silver, L. M. (2007). *Remaking Eden: How genetic engineering and cloning will transform the American family.* New York, NY: Harper Perennial.
3. For more on identifying litterers from their trace DNA and dog owners from the DNA in their dog's feces, see Mohdin, A. (2015, May 23). New campaign profiles litterers from their DNA and posts reconstructions of their faces. IFLScience. Retrieved from http://www.iflscience.com/environment/could-you-be-next-face-litter; Lacitis, E. (2015, April 3). Dog-poop DNA tests nail non-scoopers. *Seattle Times.* Retrieved from http://www.seattletimes.com/seattle-news/dog-poop-dna-tests-nail-non-scoopers/.
4. Of course, if the program's producers were serious about identifying whether this DNA sample came from Elvis, they should have genotyped a biological relative of Elvis's, and the answer would be clear.
5. There are a number of genetic variants along the MYBPC3 gene that are strongly associated with familial hypertrophic cardiomyopathy, which would be a rather deterministic cause of the disease that Dr. Kingsmore argued killed Elvis. For example, there is an autosomal dominant mutation on the MYBPC3 gene that is strongly associated with familial hypertrophic cardiomyopathy. See Dhandapany, P. S., Sadayappan, S., Xue, Y., Powell, G. T., Rani, D. S., Nallari, P., et al. (2009). A common MYBPC3 (cardiac myosin binding protein C) variant associated with cardiomyopathies in South Asia. *Nature Genetics, 41,* 187–191. However, none of the pathogenic variants on the MYBPC3 gene are the one that Elvis possessed. The variant of Elvis's that the show discussed, having a G nucleotide at RSID 193922380, has not at all been shown to be a strong predictor of familial hypertrophic cardiomyopathy and is currently very

poorly understood. It has been discussed in only a few medical papers, in which a large collection of different genetic markers along the same gene were investigated, and the competing papers described Elvis's particular variant to be "of uncertain significance" and more recently as "likely benign" with respect to familial hypertrophic cardiomyopathy. See Rodriguez-Garcia, M. I., Monserrat, L., Ortiz, M., Fernández, X., Cazón, L., Núñez, L., et al. (2010, April 30). Screening mutations in myosin binding protein C3 gene in a cohort of patients with hypertrophic cardiomyopathy. *BMC Medical Genetics*, *11*, 67. doi: 10.1186/1471-2350-11-67; Ehlermann, P., Weichenhan, D., Zehelein, J., Steen, H., Pribe, R., Zeller, R., et al. (2008, October 28). Adverse events in families with hypertrophic or dilated cardiomyopathy and mutations in the MYBPC3 gene. *BMC Medical Genetics*, *9*, 95. doi: 10.1186/1471-2350-9-95; Olivotto, I., Kassem, H., & Girolami, F. (2011). Microvascular function is selectively impaired in patients with hypertrophic cardiomyopathy and sarcomere myofilament gene mutations. *Journal of the American College of Cardiology*, *58*, 839–848; http://www.ncbi.nlm.nih.gov/clinvar/?term=rs193922380; http://www.ncbi.nlm.nih.gov/clinvar/variation/36607/.

6.  http://www.king-elvis-presley.de/html/death-elvis-autopsy.html

7.  The program identified a variant on chromosome 19, but the strongest predictors lie elsewhere. Locke, A. E., Kahali, B., Berndt, S. I., Justice, A. E., Pers, T. H., Day, F. R., et al. (2015). Genetic studies of body mass index yield new insights for obesity biology. *Nature*, *518*, 197–206.

8.  The program identified a variant on chromosome 1, but there are many predictors, and more recently the reliability of all of those identified predictors has been called into question. Palotie, A., Anttila, V., Winsvold, B. S., Gormley, P., Kurth, T., Bettella, F., McMahon, G., et al. (2013). Genome-wide analysis identifies new susceptibility loci for migraine. *Nature Genetics*, *45*, 912–916; De Vries, B., Anttila, V., Freilinger, T., Wessman, M., Kaunisto, M. A., Kallela, M., et al. (in press). Systematic re-evaluation of genes from candidate gene association studies in migraine using a large genome-wide association data set. *Cephalalagia*.

9.  The program identified a variant on chromosome 17, but among the many predictive variants, the stronger predictors are on different chromosomes. See Bailey, J. N. C., Loomis, S. J., Kang, J. H., Allingham, R. R., Gharahkhani, P., Khor, C. C., et al. (2015). Genome-wide association analysis identifies TXNRD2, ATXN2, and FOXC1 as

susceptibility loci for primary open-angle glaucoma. *Nature Genetics*, *48*, 189–194.

10. Guralnick, P. (2000). *The unmaking of Elvis Presley*. New York, NY: Back Bay Books.

11. http://www.king-elvis-presley.de/html/death-elvis-autopsy.html.

12. http://blogs.telegraph.co.uk/news/andrewmcfbrown/100012805/did-elvis-really-eat-more-than-an-asian-elephant-does/.

13. Dar-Nimrod, I., Cheung, B. Y., Ruby, M. B., & Heine, S. J. (2014). Can merely learning about obesity genes lead to weight gain? *Appetite*, *81*, 269–276.

14. See Shear, M. D. (2015, June 22). Making a point, Obama invokes a painful slur. *New York Times*. Retrieved from http://www.nytimes .com/2015/06/23/us/obama-racism-marc-maron-podcast.html?hp& action=click&pgtype=Homepage&module=first-column-region& region=top-news&WT.nav=top-news&_r=0.

15. http://www.azlyrics.com/lyrics/pink/badinfluence.html

16. http://thecelebritycafe.com/feature/2012/09/brad-pitt-talks-gun-control -says-its-our-dna-americans-own-one.

17. Quote from CNN interview, February 11, 2010. Retrieved from http://www.cnn.com/2010/SHOWBIZ/02/11/donald.trump.marriage .apprentice/.

18. http://blogs.ft.com/businessblog/2013/04/corporate-dna-jargon-is-nothing-to-celebrate/.

19. https://www.genome.gov/10002329.

20. For example, the entire founding editorial board of the American journal *Genetics* endorsed the eugenics movement. See Paul D. B. (1995). *Controlling human heredity: 1865 to the present*. Atlantic Highlands, NJ: Humanities Press. For more discussion on the links between early research on human genetics and eugenics, see Chapter 7.

21. Speech retrieved from http://www.slate.com/blogs/the_slatest/2013/ 10/29/rand_paul_gattaca_did_rachel_maddow_catch_rand_paul_ plagiarizing_wikipedia.html.

22. For example, see Metzl, J. F. (2014, October 10). The genetics epidemic: The revolution in DNA science and what to do about it. *Foreign Affairs*. Retrieved from https://www.foreignaffairs.com/articles/united -states/2014-10-10/genetics-epidemic; Kahn, Jennifer (2015, November 9). The Crispr quandary. *New York Times*. Retrieved from http://www .nytimes.com/2015/11/15/magazine/the-crispr-quandary.html? action=click&pgtype=Homepage&region=CColumn&module=Most Emailed&version=Full&src=me&WT.nav=MostEmailed&_r=0.

## 2. How Genes Make You Who You Are

1.  For information on Mendel's life, see Mawer, S. (2006). *Gregor Mendel: Planting the seeds of genetics.* New York, NY: Abrams; Henig, R. M. (2000). *The monk in the garden.* New York, NY: Mariner; Orel, V. (1996). *Gregor Mendel: The first geneticist.* Oxford: Oxford University Press.

2.  http://www.nature.com/nature/journal/v485/n7400/full/nature11119 .html.

3.  Wu, Q., Zhang, T., Cheng, J.-F., Kim, Y., Grimwood, J. H., Schmutz, J., et al. (2001). Comparative DNA sequence analysis of mouse and human protocaherin gene clusters. *Genome Research, 11,* 389–404.

4.  Welle, S. (1999). *Human protein metabolism.* New York, NY: Springer.

5.  http://www.nature.com/nature/journal/v488/n7412/full/488467a.html?WT .ec_id=NATURE-20120823.

6.  Lamason, R. L., Mohideen, M. A., Mest, J. R., Wong, A. C., Norton, H. L., Aros, M. C., et al. (2005). SLC24A5, a putative cation exchanger, affects pigmentation in zebrafish and humans. *Science, 310,* 1782–1786.

7.  Dias, B. G., & Ressler, K. J. (2014). Parental olfactory experience influences behavior and neural structure in subsequent generations. *Nature Neuroscience, 17,* 89–96.

8.  Galton, F. (1875). The history of twins, as a criterion of the relative powers of nature and nurture. *Fraser's Magazine, 12,* 566–576.

9.  Visscher, P. M. (2008). Sizing up human height variation. *Nature Genetics, 40,* 489–490.

10. Simonson, I., & Sela, A. (2011). On the heritability of consumer decision making: An exploratory approach for studying genetic effects on judgment and choice. *Journal of Consumer Research, 37,* 951–966; Plomin, R., Corley, R., DeFries, J. C., & Fulker, D. W. (1990). Individual differences in television viewing in early childhood: Nature as well as nurture. *Psychological Science, 1,* 371–377; Bouchard, T. J. (2004). Genetic influence on human psychological traits: A survey. *Current Directions in Psychological Science, 13,* 148–151; Eaves, L. J., Eysenck, H. J., & Martin, N. G. (1989). *Genes, culture and personality: An empirical approach.* San Diego, CA: Academic Press; Martin, N. G., Eaves, L. J., Heath, A. R., Jardine, R., Feingold, L. M., & Eysenck, H. J. (1986). Transmission of social attitudes. *Proceedings of the National Academy of Science, 83,* 4364–4368; Eley, T. C., Lichtenstein, P., & Stevenson, J. (1999). Sex differences in the etiology of aggressive and nonaggressive antisocial behavior: Results from two twin studies. *Child Development, 70,* 155–168; Iervolino, A. C., Perroud, N., Fullana, M. A., Guipponi, M., Cherkas,

L., Collier, D. A., et al. (2009). Prevalence and heritability of compulsive hoarding: A twin study. *American Journal of Psychiatry, 166,* 1156–1161; Pedersen, O. B., Axel, S., Rostgaard, K., Erikstrup, C., Edgren, G., Nielsen, K. R., et al. (2015). The heritability of blood donation: A population-based nationwide twin study. *Transfusion, 55,* 2169–2174.

11. Kendler, K. S., & Karkowski-Shuman, L. (1997). Stressful life events and genetic liability to major depression: Genetic control of exposure to the environment? *Psychological Medicine, 27,* 539–547.

12. But for a general critique of behavioral genetics findings, see Charney, E. (2012). Behavior genetics and postgenomics. *Behavioral and Brain Sciences, 35,* 331–358.

13. Turkheimer, E. (2000). Three laws of behavior genetics and what they mean. *Current Directions in Psychological Science, 5,* 160–164.

14. Turkheimer, E. (1998). Heritability and biological explanation. *Psychological Review, 105,* 782–791.

15. Polderman, T. J. C., Benyamin, B., de Leeuw, C. A., Sullivan, P. F., van Bochoven, A., Visscher, P. M., et al. (2015). Meta-analysis of the heritability of human traits based on fifty years of twin studies. *Nature Genetics.* doi: 10.1038/ng.3285.

16. Hugot, J.-P., Chamaillard, M., Zouali, H., Lesage, S., Cézard, J. P., Belaiche, J., et al. (2001). Association of NOD2 leucine-rich repeat variants with susceptibility to Crohn's disease. *Nature, 411,* 599–603.

17. Franke, A., McGovern, D. P., Barrett, J. C., Wang, K., Radford-Smith, G. L., Ahmad, T., et al. (2010). Genome-wide meta-analysis increases to 71 the number of confirmed Crohn's disease susceptibility loci. *Nature Genetics, 42,* 1118–1125.

18. Worden, G. (2002). *Mütter Museum of the College of Physicians of Philadelphia.* New York, NY: Blast Books.

19. Song, G-A., Kim, H-J., Woo, K-M., Baek, J-H., Kim, G-S., Choi, J-Y., & Ryoo, H-M. (2010). Molecular consequences of the ACVR1[R206H] mutation of fibrodysplasia ossificans progressive. *Journal of Biological Chemistry, 285,* 22542–22553.

20. Day, N., & Holmes, L. B. (1973). The incidence of genetic disease in a university hospital population. *American Journal of Human Genetics, 25,* 237–246.

21. Alas, even the ways that monogenic diseases emerge can be more complicated than the straightforward notion that a single gene causes the disease. For some examples, see Badano, J. L., & Katsanis, N. (2002). Beyond Mendel: An evolving view of human genetic disease transmission. *Nature Reviews Genetics, 3,* 779–789.

22. For a very thoughtful treatment of the differences between strong and weak genetic influence, see Turkheimer, E. (1998). Heritability and biological explanation. *Psychological Review, 105*, 782–791.

23. Chabris, C. F., Lee, J. J., Cesarini, D., Benjamin, D. J., & Laibson, D. I. (2015). The fourth law of behavioral genetics. *Current Directions in Psychological Science, 24*, 304–312.

24. Dobbs, D. (2015, May 21). What is your DNA worth? *BuzzFeed*. Retrieved from http://www.buzzfeed.com/daviddobbs/weighing-the-promises-of-big-genomics#.vjnnjJzwK.

25. Visscher, P. M. (2008). Sizing up human height variation. *Nature Genetics, 40*, 489–490.

26. McEvoy, B. P., & Visscher, P. M. (2009). Genetics of human height. *Economics & Human Biology, 7*, 294–306.

27. See Weedon, M. N., Lettre, G., Freathy, R. M., Lindgren, C. M., Voight, B. F., Perry, J. R., et al. (2007). A common variant of HMGA2 is associated with adult and childhood height in the general population. *Nature Genetics, 39*, 1245–1250.

28. Goldstein, D. B. (2009). Common genetic variation and human traits. *New England Journal of Medicine, 360*(17), 1696–1698.

29. Yang, J., Benyamin, B., McEvoy, B. P., Gordon, S., Henders, A. K., Nyholt, D. R., et al. (2010). Common SNPs explain a large proportion of the heritability for human height. *Nature Genetics, 42*, 565–569. Another study finds that 697 variants can explain 20 percent of the heritability for adult height: Wood, A. R., Esko, T., Yang, J., Vedantam, S., Pers, T. H., Gustafsson, S., et al. (2014). Defining the role of common variation in the genomic and biological architecture of adult human height. *Nature Genetics, 46*, 1173–1186.

30. Johnson, W. (2010). Understanding the genetics of intelligence: Can height help? Can corn oil? *Current Directions in Psychological Science, 19*, 177–182.

31. One argument for taller Dutch genes is some evidence that taller Dutch men (but not women) have more offspring, unlike Americans, where shorter men have more offspring. See Stulp, G., Barrett, L., Tropf, F. C., & Mills, M. (2015). Does natural selection favour taller stature among the tallest people on earth? *Proceedings of the Royal Society B, 282*, 20150211. doi: 10.1098/rspb.2015.0211.

32. Komlos, J., & Lauderdale, B. E. (2007). The mysterious trend in human heights in the 20th century. *American Journal of Human Biology, 34*, 206–215.

33. Greulich, W. W. (1957). A comparison of the physical growth and

development of American-born and native Japanese children. *American Journal of Physical Anthropology, 15,* 489–515.

34. Komlos, J., & Breitfelder, A. (2007). Are Americans shorter (partly) because they are fatter? A comparison of US and Dutch children's height and BMI values. *Annals of Human Biology, 34,* 593–606.

35. Wiley, A. S. (2005). Does milk make children grow? Relationships between milk consumption and height in NHANES 1999–2002. *American Journal of Human Biology, 17,* 425–441.

36. By my calculations, the correlations between national male height (there's more reliable data for males) and national milk consumption within continents range from .3 to .7, while globally across nations the correlation is .82. See data on international comparisons of height for males in http://en.wikipedia.org/wiki/Template:Average_height_around_the_world, and of annual milk consumption from http://chartsbin.com/view/1491.

37. Funatogawa, I., Funatogawa, T., Nakao, M., Karita, K., & Yano, E. (2009). Changes in body mass index by birth cohort in Japanese adults: Results from the National Nutrition survey of Japan 1956–2005. *International Journal of Epidemiology, 38,* 83–92.

38. Takahashi, E. (1984). Secular trend in milk consumption and growth in Japan. *Human Biology, 56,* 427–437.

39. Bogin, B. (1999). *Patterns of human growth* (2nd ed.). Cambridge: Cambridge University Press.

40. Lamason, R. L., Mohideen, M. A., Mest, J. R., Wong, A. C., Norton, H. L., Aros, M. C., et al. (2005). SlC24A5, a putative cation exchanger, affects pigmentation in zebrafish and humans. *Science, 310,* 1782–1786.

41. Some good examples of the ways that people avoid fully understanding something and taking the easy way out can be found in Petty, R. E., & Cacioppo, J. T. (1984). The effects of involvement on responses to argument quantity and quality: Central and peripheral routes to persuasion. *Journal of Personality and Social Psychology, 46,* 69–81; and Ratneshwar, S., & Chaiken, S. (1991). Comprehension's role in persuasion: The case of its moderating effect on the persuasive impact of source cues. *Journal of Consumer Research, 18,* 52–62.

42. Jablonka, E., & Lamb, M. J. (2006). *Evolution in four dimensions: Genetic, epigenetic, behavioral, and symbolic variation in the history of life.* Cambridge, MA: MIT Press. Also see Strohman, R. (1995). Linear genetics, non-linear epigenetics: Complementary approaches to understanding complex diseases. *Integrative Physiological and Behavioral Science, 30*(4), 273–282.

43. Sturtevant, A. H. (1965). *A history of genetics*. New York, NY: Harper and Row.

44. Mendel was indeed lucky with the results from his studies. The data from his pea studies had such a strikingly clear pattern that some of his critics have accused him of doctoring his data. Ronald Fisher, the famed statistician and geneticist, is the most dogged of those who questioned the fidelity of Mendel's findings. See Fisher, R. A. (1936). Has Mendel's work been rediscovered? *Annals of Science, 1*, 115–137.

45. Henig, R. M. (2000). *The monk in the garden*. New York, NY: Mariner.

## 3. Your Genes, Your Soul?

1. See p. 51 of Lombroso, C. (2006). *Criminal Man* (M. Gibson & N. H. Rafter, Trans.). Durham, NC: Duke University Press. This edition contains a new translation with an introduction and notes by the translators, Mary Gibson and Nicole Hahn Rafter.

2. Macnamara, J. (1986). *A border dispute*. Cambridge, MA: MIT Press.

3. Locke, J. (1671/1959). *An essay concerning human understanding, Vol. 2.* New York, NY: Dover.

4. See Medin, D. L., & Ortony, A. (1989). Psychological essentialism. In S. Vosniadou & A. Ortony (Eds.), *Similarity and analogical reasoning* (pp. 179–195). New York, NY: Cambridge University Press.

5. For a thorough review of essentialism, see Gelman, S. A. (2003). *The essential child: Origins of essentialism in everyday thought*. New York, NY: Oxford University Press.

6. Nemeroff, C., & Rozin, P. (1989). You are what you eat: Applying the demand-free "impressions" technique to an unacknowledged belief. *Ethos, 17*, 50–69.

7. See Goldman, L. R. (Ed.). (1999). *The anthropology of cannibalism*. Santa Barbara, CA: Praeger.

8. I'm indebted to Francisco Gil-White for his descriptions of essentialist thinking. For more discussion on the role of essences in natural kinds, see Gil-White, F. J. (2001). Are ethnic groups biological "species" to the human brain? Essentialism in our cognition of some social categories. *Current Anthropology, 42*, 515–554.

9. Ereshefsky, M. (2004). *The poverty of the Linnaean hierarchy*. New York, NY: Cambridge University Press.

10. http://www.dailymail.co.uk/health/article-558256/I-given-young-mans

-heart---started-craving-beer-Kentucky-Fried-Chicken-My-daughter -said-I-walked-like-man.html.

11. Sylvia, C., & Novak, W. (1997). *A change of heart*. New York, NY: Warner Books.

12. Inspector, I., Kutz, Y., & David, D. (2004). Another person's heart: Magical and rational thinking in the psychological adaptation to heart transplantation. *The Israeli Journal of Psychiatry and Related Sciences*, *41*, 161–173.

13. Hood, B. M., Gjersoe, N. L., Donnelly, K., Byers, A., & Itakura, S. (2011). Moral contagion attitudes towards potential organ transplants in British and Japanese adults. *Journal of Cognition and Culture*, *11*, 269–286.

14. http://www.dailymail.co.uk/news/article-557864/Man-given-heart -suicide-victim-marries-donors-widow-kills-exactly-way.html.

15. Nemeroff, C., & Rozin, P. (1994). The contagion concept in adult thinking in the United States: Transmission of germs of interpersonal influence. *Ethos, 22*, 158–186.

16. Because psychologists have typically assumed that most psychological phenomena are universal, they have rarely pursued the more costly and inconvenient research necessary to explore the cultural boundaries of various ways of thinking. Most psychological research is conducted with convenience samples, which tend to be American college students who are enrolled in psychology classes. Indeed, American college students are more than 4,000 times more likely to participate in a psychology experiment compared with adults from non-Western countries. Psychology is curiously dominated by American research as well as research from other English-speaking countries, such as the UK, Canada, and Australia. So for many, if not most, psychological phenomena, we really don't have a great understanding of how similar or different the ways of thinking appear around the world. The study of essentialism, in contrast, has very much been a global enterprise.

Henrich, J., Heine, S. J., & Norenzayan, A. (2010). The weirdest people in the world. *Behavioral and Brain Sciences, 33*, 61–83.

17. Gil-White, F. J. (2001). Are ethnic groups biological "species" to the human brain? Essentialism in our cognition of some social categories. *Current Anthropology, 42*, 515–554.

18. Much research finds that people's preferences for essentialist thinking are not quite as strong among people from a variety of Asian societies, such as China, Japan, and India (although it's still clearly evident there

too), as they are among North Americans. See, for example, Choi, I., Nisbett, R. E., & Norenzayan, A. (1999). Causal attribution across cultures: Variation and universality. *Psychological Bulletin, 125,* 47–63; Rattan, A., Savani, K. S., Naidu, N. V. R., & Dweck, C. S. (2012). Can everyone become highly intelligent? Cultural differences in and societal consequences of beliefs about the universal potential for intelligence. *Journal of Personality and Social Psychology, 103,* 787–802; and Heine, S. J., Kitayama, S., Lehman, D. R., Takata, T., Ide, E., Leung, C., & Matsumoto, H. (2001). Divergent consequences of success and failure in Japan and North America: An investigation of self-improving motivations and malleable selves. *Journal of Personality and Social Psychology, 81,* 599–615.

19. Kraus, M. W., & Keltner, D. (2013). Social class rank, essentialism, and punitive judgment. *Journal of Personality and Social Psychology, 105,* 247–261. Also see Kraus, M. W., Piff, P. K., Mendoza-Denton, R., Rheinschmidt, M. L., & Kelter, D. (2012). Social class, solipsism, and contextualism: How the rich are different from the poor. *Psychological Review, 119,* 546–572.

20. Mahalingam, R. (1998). *Essentialism, power, and representation of caste: A developmental study.* Ph.D. dissertation, University of Pittsburgh.

21. Keil, F. C. (1989). *Concepts, kinds, and cognitive development.* Cambridge, MA: MIT Press.

22. Gelman, S. A., Frazier, B. N., Noles, N. S., Manczak, E. M., & Stilwell, S. M. (2015). How much are Harry Potter's glasses worth? Children's monetary evaluation of authentic objects. *Journal of Cognition and Development, 16,* 97–117.

23. Atran, S. (1998). Folk biology and the anthropology of science: Cognitive universals and cultural particulars. *Behavioral and Brain Sciences, 21,* 547–69. Also see Gil-White, F. J. (2001). Are ethnic groups biological "species" to the human brain? Essentialism in our cognition of some social categories. *Current Anthropology, 42,* 515–554.

24. Medin, D. L., & Ortony, A. (1989). Psychological essentialism. In S. Vosniadou & A. Ortony (Eds.), *Similarity and analogical reasoning* (pp. 179–195). New York, NY: Cambridge University Press.

25. Quoted in Jaoff, Leon J. (1989, March 20). The gene hunt, *Time,* 62–67.

26. Quoted in the *New York Times* article Degrees offered in genetic counseling, December 6, 1970, p. 71.

27. Lanie, A. D., Jayaratne, T. E., Sheldon, J. P., Kardia, S. L. R., Anderson, E. S., Feldbaum, M., & Petty, E. M. (2004). Exploring the public

understanding of basic genetic concepts. *Journal of Genetic Counseling*, *13*, 305–320.

28. Heyman, G. D., & Gelman, S. A. (2000). Beliefs about the origins of human psychological traits. *Developmental Psychology*, *36*, 663–678.

29. Dar-Nimrod, I., & Heine, S. J. (2011). Genetic essentialism: On the deceptive determinism of DNA. *Psychological Bulletin*, *137*, 800–818.

30. See, for example, http://www.cbc.ca/day6/blog/2010/12/10/comment-the-infidelity-gene/.

31. Garcia, J. R., MacKillop, J., Aller, E. L., Merriwether, A. M., Wilson, D. S., & Lum, J. K. (2010). Associations between dopamine D4 receptor gene variation with both infidelity and sexual promiscuity. *PLoS ONE* *5*(11), e14162. doi: 10.1371/journal.pone.0014162.

32. See Frankena, W. K. (1939). The naturalistic fallacy. *Mind*, *48*, 464–477; Moore, G. E. (1903). *Principia ethica*. New York, NY: Cambridge University Press.

33. Scott, S. E., Inbar, Y., & Rozin, P. (in press). Evidence for absolute moral opposition to genetically modified food in the United States. *Perspectives in Psychological Science*.

## 4. The Oracle at 23andMe: Genetic Testing and Disease

1. Becker, E. (1973). *The denial of death*. New York, NY: Free Press.

2. http://www.gpo.gov/fdsys/pkg/CHRG-111shrg47375/html/CHRG-111shrg47375.htm; http://www.fda.gov/downloads/AdvisoryCommittees/CommitteesMeetingMaterials/MedicalDevices/MedicalDevicesAdvisory Committee/MolecularandClinicalGeneticsPanel/UCM248564.ppt. For a nice description of Gulcher's story, see Davies, K. (2010). *The $1000 genome*. New York, NY: Free Press.

3. http://www.nytimes.com/2013/05/14/opinion/my-medical-choice .html?8qa&_r=0&gwh=4A9AB8DDD848F5C8702A7A1109360F8E& gwt=pay&assetType=opinion.

4. Dar-Nimrod, I., Cheung, B. Y., Ruby, M. B., & Heine, S. J. (2014). Can merely learning about obesity genes lead to weight gain? *Appetite*, *81*, 269–276.

5. Caspi, A., Sugden, K., Moffitt, T. E., Taylor, A., Craig, I. W., Harrington, H., McClay, J., et al. (2003). Influence of life stress on depression: Moderation by a polymorphism in the 5-HTT gene. *Science*, *301*, 386–389.

6. Schomerus, G., Schwahn, C., Holzinger, A., Corrigan, P. W., Grabe,

H. J., Carta, M. G., et al. (2012). Evolution of public attitudes about mental illness: A systematic review and meta-analysis. *Acta Psychiatrica Scandinavica, 125,* 440–452; Pescosolido, B. A., Martin, J. K., Long, J. S., Medina, T. R., Phelan, J. C., & Link, B. G. (2010). "A disease like any other?" A decade of change in public reactions to schizophrenia, depression, and alcohol dependence. *American Journal of Psychiatry, 167,* 1321–1330.

7. Lebowitz, M. S., & Ahn, W.-K. (2014). Effects of biological explanations for mental disorders on clinicians' empathy. *Proceedings of the National Academy of Sciences, 111,* 17786–17790; Phelan, J. C., Yang, L. H., & Cruz-Rojas, Rr. (2006). Effects of attributing serious mental illnesses to genetic causes on orientations to treatment. *Psychiatric Services, 57,* 382–387.

8. Senior, V., Marteau, T. M., & Weinman, J. (2000). Impact of genetic testing on causal models of heart disease and arthritis: An analogue study. *Psychology & Health, 14*(6), 1077–1088; Phelan, J., Cruz-Rojas, R., & Reiff, M. (2002). Genes and stigma: The connection between perceived genetic etiology and attitudes and beliefs about mental illness. *Psychiatric Rehabilitation Skills, 6,* 159–185; Deacon, B. J., & Baird, G. L. (2009). The chemical imbalance explanation of depression: Reducing blame at what cost? *Journal of Social and Clinical Psychology, 28,* 415–535; Goldstein, B., & Rosselli, F. (2003). Etiological paradigms of depression: The relationship between perceived causes, empowerment, treatment preferences, and stigma. *Journal of Mental Health, 12,* 551–563. Although note that there's some evidence that clinicians don't become more empathic when considering biological explanations: Lebowitz, M. S., & Ahn, W.-K. (2014). Effects of biological explanations for mental disorders on clinicians' empathy. *Proceedings of the National Academy of Sciences, 111,* 17786–17790.

9. Haslam, N., & Kvaale, E. P. (2015). Biogenetic explanations of mental disorder: The mixed-blessings model. *Current Directions in Psychological Science, 24,* 399–404.

   Kvaale, E. P., Haslam, N., & Gottdiener, W. H. (2013). The "side effects" of medicalization: A meta-analytic review of how biogenetic explanations affect stigma. *Clinical Psychology Review, 33,* 782–794.

10. Link, B. G., & Phelan, J. C. (2006). Stigma and its public health implications. *Lancet, 367,* 528–529.

11. Mehta, S., & Farina, A. (1997). Is being "sick" really better? Effect of the disease view of mental disorder on stigma. *Journal of Social & Clinical Psychology, 16,* 405–419.

12. Kessler, R. C., McGonagle, K. A., Zhao, S., Nelson, C. B., Hughes, N., Eshleman, S., Wittchen, H.-U., & Kendler, K. S. (1994). Lifetime and 12-month prevalence of *DSM-III*-R psychiatric disorders in the United States: Results from the National Comorbidity Survey. *Archives of General Psychiatry, 51,* 8–19.

13. Teasdale, J. D. (1983). Negative thinking in depression: Cause, effect, or reciprocal relationship? *Advances in Behaviour Research and Therapy, 5,* 3–25.

14. Kvaale, E. P., Haslam, N., & Gottdiener, W. H. (2013). The "side effects" of medicalization: A meta-analytic review of how biogenetic explanations affect stigma. *Clinical Psychology Review, 33,* 782–794; Lebowitz, M. S., Ahn, W. K., & Nolen-Hoeksema, S. (2013). Fixable or fate? Perceptions of the biology of depression. *Journal of Consulting and Clinical Psychology, 81,* 518–527.

15. Leamy, M., Bird, V., Le Boutillier, C., Williams, J., & Slade, M. (2011). Conceptual framework for personal recovery in mental health: Systematic review and narrative synthesis. *The British Journal of Psychiatry, 199,* 445–452; Carver, C. S., Scheier, M. F., & Segerstrom, S. C. (2010). Optimism. *Clinical Psychology Review, 30,* 879–889.

16. For a long review of the many effects that have been linked to the 5-HTTLPR gene, see Table 2 in Charney, E., & English, E. (2012). Candidate genes and political behavior. *American Political Science Review, 106,* 1–34.

17. Risch, N., Herrell, R., Lehner, T., Liang, K., Eaves, L., Hoh, J., et al. (2009). Interaction between the serotonin transporter gene (5-HTTLPR), stressful life events and risk of depression: A meta-analysis. *Journal of the Medical Association of America, 301,* 2462–2471.

18. Kaufman, D. J., Bollinger, J. M., Dvoskin, R. L., & Scott, J. A. (2012). Risky business: Risk perception and the use of medical services among customers of DTC personal genetic testing. *Journal of Genetic Counseling, 21,* 413–422.

19. Campeau, P. M., Foulkes, W. D., & Tischkowitz, M. D. (2008). Hereditary breast cancer: New genetic developments, new therapeutic avenues. *Human Genetics, 124,* 31–42.

20. Basback, N., Sizto, S., Guh, D., & Anis, A. H. (2012). The effect of direct-to-consumer genetic tests on anticipated affect and health-seeking behaviors: A pilot survey. *Genetic Testing and Molecular Biomarkers, 16,* 1165–1171.

21. Long, C. (2010, March 14). When DNA means do not ask. *The Sunday Times.*

22. Gordon, E. S., Griffin, G., Wawak, L., Pang, H., Gollust, S. E., & Bernhardt, B. A. (2012). "It's not like judgment day": Public understanding and reactions to personalized genomic risk information. *Journal of Genetic Counselling, 21*, 423–432.

23. Baumeister, R. F., Bratslavsky, E., Finkenauer, C., & Vohs, K. D. (2001). Bad is stronger than good. *Review of General Psychology, 5*, 323–370.

24. See Rozin, P., & Royzman, E. B. (2001). Negativity bias, negativity dominance, and contagion. *Personality and Social Psychology Review, 5*, 296–320.

25. Lerman, C., Croyle, R. T., Tercyak, K. P., & Hamann, H. (2002). Genetic testing: Psychological aspects and implications. *Journal of Consulting and Clinical Psychology, 3*, 784–797.

26. Bloss, C. S., Schork, N. J., & Topol, E. J. (2011). Effect of direct-to-consumer genomewide profiling to assess disease risk. *The New England Journal of Medicine, 364*, 524–534; Bloss, C. S., Wineinger, N. E., Darst, B. F., Schork, N. J., & Topol, E. J. (2013). Impact of direct-to-consumer genomic testing at long term follow-up. *Journal of Medical Genetics, 50*, 393–400; Egglestone, C., Morris, A., & O'Brien, A. (2013). Effect of direct-to-consumer genetic tests on health behavior and anxiety: A survey of consumers and potential consumers. *Journal of Genetic Counselling, 22*, 565–575; Gordon, E. S., et al. (2012). "It's not like judgment day": Public understanding and reactions to personalized genomic risk information. *Journal of Genetic Counselling, 21*, 423–432; Hershka, J. T., Palleschi, C., Howley, H., Wilson, B., & Wells, P. S. (2008). A systematic review of perceived risks, psychological and behavioral impacts of genetic testing. *Genetics in Medicine, 10*, 19–32.

27. Bloss, C. S., Schork, N. J., & Topol, E. J. (2011). Effect of direct-to-consumer genomewide profiling to assess disease risk. *The New England Journal of Medicine, 364*, 524–534.

28. Miller, D. T. (2010). Psychological impact of genetic testing. In T. W. Miller (Ed.), Handbook of stressful transitions across the lifespan (pp. 585–604). New York, NY: Springer Science; Cheung, B. Y., Dar-Nimrod, I., & Gonsalkorale, K. (2014). Am I my genes? Perceived genetic etiology, intrapersonal processes, and health. *Social and Personality Psychology Compass, 8*(11), 626–637; Van Oostrom, I., Meijers-Heijboer, H., Duivenvoorden, H. J., Bröcker-Vriends, A. H. J. T., Van Asperen, C. J., Sijmons, R. H., et al. (2007). Prognostic factors for hereditary cancer distress six months after BRCA1/2 or HNPCC genetic susceptibility testing. *European Journal of Cancer, 43*, 71–77.

29. Wilson, T. D., Wheatley, T., Meyers, J. M., Gilbert, D. T., & Axsom,

D. (2000). Focalism: A source of durability bias in affective forecasting. *Journal of Personality and Social Psychology, 78*, 821–836.

30. Peters, S. A., Laham, S. M., Pachter, N., & Winsip, L. M. (2014). The future in clinical genetics: Affective forecasting biases in patient and clinician decision making. *Clinical Genetics, 85*, 312–317.

31. Brickman, P., & Campbell, D. T. (1971). Hedonic relativism and planning the good society. In H. M. Appley (Ed.), *Adaptation level theory: A symposium*. New York: Academic Press.

32. For an excellent review on this, see Wilson, T. D., & Gilbert, D. T. (2008). Explaining away: A model of affective adaptation. *Perspectives on Psychological Science, 3*, 370–386.

33. Klein, J. (1980). *Woody Guthrie: A life*. New York, NY: Knopf.

34. Langbehn, D. R., Brinkman, R. R., Falush, D., Paulsen, J. S., & Hayden, M. R. (2004). A new model for prediction of the age of onset and penetrance for Huntington's disease based on CAG length. *Clinical Genetics, 65*, 267–277.

35. Wexler, N. (1992). Clairvoyance and caution: Repercussions from the Human Genome Project. In D. J. Kevles & L. Hood (Eds.), *The code of codes* (pp. 211–243). Cambridge, MA: Harvard University Press.
Revkin, A. (1993, December). Hunting down Huntington's. *Discover Magazine*.

36. Lerman, C., Croyle, R. T., Tercyak, K. P., & Hamann, H. (2002). Genetic testing: Psychological aspects and implications. *Journal of Consulting and Clinical Psychology, 3*, 784–797.

37. http://www.cbsnews.com/stories/2004/11/18/eveningnews/main656527.shtml?CMP=ILC-SearchStories.

38. Van der Steenstraten, I. M., Tibben, A., Roos, R. A., van de Kamp, J. J., & Niermeijer, M. F. (1994). Predictive testing for Huntington disease: Nonparticipants compared with participants in the Dutch program. *American Journal of Human Genetics, 55*, 618–625.

39. Wiggins, S., Whyte, P., Huggins, M., Adam, S., Theilmann, J., Bloch, M., et al. (1992). The psychological consequences of predictive testing for Huntington's disease. *The New England Journal of Medicine, 327*, 1401–1405.

40. Hoffrage, U., Lindsey, S., Hertwig, R., & Gigerenzer, G. (2000). Communicating statistical information. *Science, 290*, 2261–2262.

41. Gordon, E. S., et al. (2012). "It's not like judgment day": Public understanding of and reactions to personalized genomic risk information. *Journal of Genetic Counseling, 21*, 423–432.

42. Wang, C., Gonzalez, R., & Merajver, S. D. (2004). Assessment of genetic

testing and related counseling services: Current research and future directions. *Social Science & Medicine, 58,* 1427–1442.

43. Press, N. A., Yasui, Y., Reynolds, S., Durfy, S. J., & Burke W. (2001). Women's interest in genetic testing for breast cancer susceptibility may be based on unrealistic expectations. *American Journal of Medical Genetics, 99,* 99–110.

44. Keller, M. F., Saad, M., Bras, J., Bettella, F., Nicolaou, N., Simón-Sánchez, J., et al. (2012). Using genome-wide complex trait analysis to quantify "missing heritability" in Parkinson's disease. *Human Molecular Genetics, 21,* 4996–5009.

45. Quoted from an interview reported in Davies, K. (2010). *The $1000 genome.* New York, NY: Free Press, p. 191.

46. https://www.23andme.com/you/journal/alzheimers/techreport/.

47. Pinker, S. (2009, January 7). My genome, my self. *New York Times Magazine, New York Times.*

48. Wilkie, A. O. M. (2001). Genetic prediction: What are the limits? *Studies in History and Philosophy of Biological and Biomedical Sciences, 32,* 619–633.

49. Gottesman, I. I. (1991). *Schizophrenia. The origins of madness.* New York: Holt.

50. The International Schizophrenia Consortium (2009). Common polygenic variation contributes to risk of schizophrenia and bipolar disorder. *Nature, 460,* 748–752.

      Geschwind, D. H., & Flint, J. (2015). Genetics and genomics of psychiatric disease. *Science, 349,* 1489–1494.

      Schizophrenic Working Group of the Psychiatric Genomics Consortium. (2014). Biological insights from 108 schizophrenia-associated genetic loci. *Nature, 511,* 421–427.

51. Wade, N. (2009, July 1). Hoopla, and disappointment, in schizophrenia research. *New York Times.*

52. Juran, B. D., & Lazaridis, K. N. (2011). Genomics in the post-GWAS era. *Seminars in Liver Disease, 31*(2), 215–222.

      Need, A. C., & Goldstein, D. B. (2010). Whole genome association studies in complex diseases: Where do we stand? *Dialogues in Clinical Neuroscience, 12,* 37–46.

53. Hunter, D. J., Khoury, M. J., & Drazen, J. M. (2008). Letting the genome out of the bottle—Will we get our wish? *New England Journal of Medicine, 358,* 105–107.

54. Wade, N. (2009, April 16). Genes show limited value in predicting diseases. *New York Times.*

55. Frebourg, T. (2012). Direct-to-consumer genetic testing services: What are the medical benefits? *European Journal of Human Genetics, 20,* 483–485.

   For a similarly critical view of genome-wide association testing for common diseases, see McClellan, J., & King, M.-C. (2010). Genetic heterogeneity in human disease. *Cell, 141,* 210–217.

56. Spiegel interview with Craig Venter. (2010, July 29). Retrieved from http://www.spiegel.de/international/world/spiegel-interview-with-craig -venter-we-have-learned-nothing-from-the-genome-a-709174-2.html.

57. Paynter, N. P., Chasman, D. I., Paré, G., Buring, J. E., Cook, N. R., Miletich, J. P., et al. (2010). Association between a literature-based genetic risk score and cardiovascular events in women. *Journal of the American Medical Association, 303,* 631–637.

58. http://www.reuters.com/article/2012/01/10/us-dna-reader-newspro -idUSTRE8090WF20120110.

59. White Paper 23-01. Estimating genotype-specific incidence for one or several loci. https://23andme.https.internapcdn.net/res/pdf/HIC -SXIYiYqXreldAxO5yA_23-01_Estimating_Genotype_Specific_ Incidence.pdf.

60. Kalf, R. R. J., Mihaescu, R., Kundu, S., de Knijff, P., Green, R. C., & Janssens, A. C. J. W. (2013). Variations in predicted risks in personal genome testing for complex diseases. *Genetics in Medicine, 16,* 85–91.

61. Murray, A. B., Carson, M. J., Morris, C. A., & Beckwith, J. (2010). Illusions of scientific legitimacy: misrepresented science in the direct-to-consumer genetic-testing marketplace. *Trends in Genetics, 26*(11), 459–461.

62. Imai, K., Kricka, L. J., & Fortina, P. (2011). Concordance study of 3 direct-to-consumer genetic-testing services. *Clinical Chemistry, 57,* 518–521.

63. http://www.nytimes.com/2013/12/31/science/i-had-my-dna-picture -taken-with-varying-results.html?hp&_r=1&.

64. Fleming, Nic. (2008, September 7). Rival genetic tests leave buyers confused. *The Sunday Times.* Retrieved from http://www.thesundaytimes .co.uk/sto/news/uk_news/article234529.ece.

65. Ng, P. C., Murray, S. S., Levy, S., & Venter, J. C. (2009). An agenda for personalized medicine. *Nature, 461,* 724–726.

   Kalf, R. R. J., Mihaescu, R., Kundu, S., de Knijff, P., Green, R. C., & Janssens, A. C. J. W. (2013). Variations in predicted risks in personal genome testing for complex diseases. *Genetics in Medicine, 16,* 85–91.

66. Adams, S. D., Evans, J. P., & Ayslworth, A. S. (2013). Direct-to-consumer

genomic testing offers little clinical utility but appears to cause minimal harm. *North Carolina Medical Journal, 74,* 494–499.

67. Kutz G. (2010). Direct-to-consumer genetic tests: misleading test results are further complicated by deceptive marketing and other questionable practices. Testimony before the Subcommittee on Oversight and Investigations, Committee on Energy and Commerce, House of Representatives. Washington, DC: US Government Accountability Office. http://www.gao.gov/assets/130/125 079.pdf.

68. Spencer, D. H., Lockwood, C., Topol, E., Evans, J. P., Green, R. C., Mansfield, E., et al. (2011). Direct-to-consumer genetic testing: Reliable or risky? *Clinical Chemistry, 57,* 1641–1644.

69. Green, R. C., & Farahany, N. A. (2014). The FDA is overcautious on consumer genomics. *Nature, 505,* 286–287.

    Dobbs, D. (2013, November 27). The FDA vs. personal genetic testing. *The New Yorker.*

70. http://gigaom.com/2013/12/02/23andme-hit-with-class-action-over
    -misleading-genetic-ads/?utm_source=feedburner&utm_medium=
    feed&utm_campaign=Feed%3A+OmMalik+%28GigaOM%3A+Tech%29

71. Vayena, E., Gourna, E., Steuli, J., Hafen, E., & Prainsack, B. (2012). Experiences of early users of direct-to-consumer genomics in Switzerland: An exploratory study. *Public Health Genomics, 15,* 352–362.

## 5. Born This Way: Thinking about Gender and Sexual Orientation

1. Gelman, S. A., & Taylor, M. G. (2000). Gender essentialism in cognitive development. In P. H. Miller & S. E. Kofsky (Eds.), *Toward a feminist developmental psychology* (pp. 169–190). Florence, KY: Taylor & Frances/Routledge; Haslam, N., Rothschild, L., & Ernst, D. (2000). Essentialist beliefs about social categories. *British Journal of Social Psychology, 39,* 113–127.

2. Gelman, S. A., Collman, P., & Maccoby, E. E. (1986). Inferring properties from categories versus inferring categories from properties: The case of gender. *Child Development, 57,* 396–404.

3. Gelman, S. A. (2003). *The essential child* (p. 97). Oxford, UK: Oxford University Press.

4. Cordua, G. D., McGraw, K. O., & Drabman, R. S. (1979). Doctor or nurse: Children's perceptions of sex typed occupations. *Child Development, 50,* 590–593.

5. Hooper, E. M. (1890, November). Hints on home dress-making. *The Ladies' Home Journal*. Retrieved from http://search.proquest.com/docview/137177426.

6. See the extensive cross-cultural study by Williams, J. E., & Best, D. L. (1990). *Sex and psyche: Gender and self viewed cross-culturally. Volume 13, Cross-Cultural Research and Methodology Series*. Newbury Park, CA: Sage.

7. For an interesting demonstration that cultural gender roles are tied to historical agricultural practices involving the use of the plough, see Alesina, A., Giuliano, P., & Nunn, N. (2013). On the origins of gender roles: Women and the plough. *The Quarterly Journal of Economics*, *128*, 469–530.

8. For example, see Fausto-Sterling, A. (1985). *Myths of gender: Biological theories about women and men*. New York, NY: Basic Books.

9. For more information about David's life, see Colapinto, J. (2000). *As nature made him*. Toronto, ON: HarperCollins; Diamond, M. N., & Sigmundson, H. K. (1997). Sex reassignment at birth: A long term review and clinical implications. *Archives of Pediatrics and Adolescent Medicine*, *151*, 298–304; http://www.bbc.co.uk/sn/tvradio/programmes/horizon/dr_money_prog_summary.shtml; http://www.slate.com/articles/health_and_science/medical_examiner/2004/06/gender_gap.html.

10. Reiner, W. G., & Gearhart, J. P. (2004). Discordant sexual identity in some genetic males with cloacal exstrophy assigned to female sex at birth. *The New England Journal of Medicine*, *350*, 333–341.

11. For a review, see Maccoby, E. E. (1998). *The two sexes: Growing up apart and coming together*. Cambridge, MA: Harvard University Press.

12. Hassett, J. M., Siebert, E. R., & Wallen, K. (2008). Sex differences in rhesus monkey toy preferences parallel those of children. *Hormones and Behavior*, *54*, 359–364.

13. Kahlenberg, S. M., & Wrangham, R. W. (2010). Sex differences in chimpanzees' use of sticks as play objects resemble those of children. *Current Biology*, *20*, R106–R1068.

14. Research employing both self-report questionnaires and implicit measures finds that prepubescent transgender children identify as much with their expressed gender as do non-transgender (cisgender) children of the same expressed gender. See Olson, K. R., Key, A. C., & Eaton, N. R. (2015). Gender cognition in transgender children. *Psychological Science*, *26*, 467–474.

15. McConahay, J. B. (1983). Modern racism and modern discrimination: The effects of race, racial attitudes, and context on simulated hiring decisions. *Personality and Social Psychology Bulletin*, *9*, 551–558; Greenwald,

A. G., & Banaji, M. R. (1995). Implicit social cognition: Attitudes, self-esteem, and stereotypes. *Psychological Review, 102,* 4–27; Payne, B. K., Cheng, C. M., Govorun, O., & Stewart, B. D. (2005). An inkblot for attitudes: Affect misattribution as implicit measurement. *Journal of Personality and Social Psychology, 89,* 277–293.

16. Norton, A. T., & Herek, G. M. (2013). Heterosexuals' attitudes toward transgender people: Findings from a national probability sample of US adults. *Sex Roles, 68,* 738–753; Grant, J. M., Mottet, L. A., Tanis, J., Harrison, J., Herman, J. L., & Keisling, M. (2011). *Injustice at every turn: A report of the national transgender discrimination survey.* Washington, DC: National Center for Transgender Equality and National Gay and Lesbian Task Force; Amnesty International. (2001). *Crimes of hate, conspiracy of silence. Torture and ill-treatment based on sexual identity.* London: Author; Lombardi, E. L., Wilchins, R. A., Priesing, D., & Malouf, D. (2001). Gender violence: Transgender experiences with violence and discrimination. *Journal of Homosexuality, 42,* 89–101.

17. http://www.cnn.com/2010/LIVING/03/10/her.name.was.steven/; http://en.wikipedia.org/wiki/Susan_Stanton.

18. Grant, J. M., Mottet, L. A., Tanis, J., Harrison, J., Herman, J. L., & Keisling, M. (2011). Injustice at every turn: A report of the National Transgender Discrimination Survey. Washington, DC: National Center for Transgender Equality and National Gay and Lesbian Task Force.

19. See p. 72 of Bornstein, K. (1995). *Gender outlaw: On men, women, and the rest of us.* New York, NY: Vintage.

20. See http://www.rushlimbaugh.com/daily/2014/02/07/lesley_stahl_shocker_men_and_women_are_different.

21. See http://www.addictinginfo.org/2012/03/08/35-hateful-and-stupid-rush-limbaugh-quotes/.

22. Keller, J. (2005). In genes we trust: The biological component of psychological essentialism and its relationship to mechanisms of motivated social cognition. *Journal of Personality and Social Psychology, 88,* 686–702.

23. Morton, T. A., Postmes, T., Haslam, S. A., & Hornsey, M. J. (2009). Theorizing gender in the face of social change: Is there anything essential about essentialism? *Journal of Personality and Social Psychology, 96,* 653–664.

24. Coleman, J., & Hong, Y. (2008). Beyond nature and nurture: The influence of lay gender theories on self-stereotyping. *Self & Identity, 7,* 34–53.

25. https://web.archive.org/web/20080130023006/http://www.president.harvard.edu/speeches/2005/nber.html.

26. For competing perspectives on the evidence of biological factors underlying sex differences in math and science performance, see Pinker, S., & Spelke, E. (2005). The science of gender and science: Pinker vs. Spelke—A debate. Retrieved from http://www.edge.org/3rd_culture/debate05/debate05_index.html; Ceci, S. J., Williams, W. M., & Barnett, S. M. (2009). Women's underrepresentation in science: Sociocultural and biological considerations. *Psychological Bulletin, 135,* 218–261; Miller, D. I., & Halpern, D. F. (2014). The new science of cognitive sex differences. *Trends in Cognitive Sciences, 18,* 37–45.

27. Steele, C. M., & Aronson, J. (1995). Stereotype threat and the intellectual test performance of African Americans. *Journal of Personality and Social Psychology, 69,* 797–811.

28. See Schmader, T., Johns, M., & Forbes, C. (2008). An integrated process model of stereotype threat effects on performance. *Psychological Review, 115,* 336–356.

29. Horton, S., Baker J., Pearce, W., & Deakin, J. (2010). Immunity to popular stereotypes of aging? Seniors and stereotype threat. *Educational Gerontology, 36,* 353–371; Stone, J., Perry, W., & Darley, J. (1997). "White men can't jump": Evidence for the perceptual confirmation of racial stereotypes following a basketball game. *Basic and Applied Social Psychology, 19,* 291–306; Spencer, S. J., Steele, C. M., & Quinn, D. M. (1998). Stereotype threat and women's math performance. *Journal of Experimental Social Psychology, 35,* 4–28.

30. Dar Nimrod, I., & Heine, S. J. (2006). Exposure to scientific theories affects women's math performance. *Science, 314,* 435.

31. Pinker, S., & Spelke, E. (2005). The science of gender and science: Pinker vs. Spelke—A debate. Retrieved from http://www.edge.org/3rd_culture/debate05/debate05_index.html; Ceci, S. J., Williams, W. M., & Barnett, S. M. (2009). Women's underrepresentation in science: Sociocultural and biological considerations. *Psychological Bulletin, 135,* 218–261; Miller, D. I., & Halpern, D. F. (2014). The new science of cognitive sex differences. *Trends in Cognitive Sciences, 18,* 37–45.

32. Moore, J., & Slater, W. (2006). *The architect: Karl Rove and the master plan for absolute power.* New York, NY: Crown.

33. Donovan, T., Tolbert, C., Smith, D. A., & Parry, J. (2005). *Did gay marriage elect George W. Bush?* Paper presented at the Western Political Science Association, March 17–20, Oakland, CA.

34. See Hegarty, P., & Pratto, F. (2001). Sexual orientation beliefs: Their relationship to antigay attitudes and biological determinist arguments. *Journal of Homosexuality, 41,* 121–135; Jayaratne, T. E., Ybarra, O.,

Sheldon, J. P., Brown, T. N., Feldbaum, M., Pfeffer, C., et al. (2006). White Americans' genetic lay theories of race differences and sexual orientation: Their relationship with prejudice toward Blacks and gay men and lesbians. *Group Processes and Intergroup Relations, 9*, 77–94; Smith, S. J., Zanotti, D. C., Axelton, A. M., & Saucier, D. A. (2011). Individuals' beliefs about the etiology of same-sex sexual orientation. *Journal of Homosexuality, 58*, 1110–1131; Tygart, C. E. (2000). Genetic causation attribution and public support of gay rights. *International Journal of Public Opinion Research, 12*, 259–275.

35. Haslam, N., & Levy, S. R. (2007). Essentialist beliefs about homosexuality: Structure and implications for prejudice. *Personality and Social Psychology Bulletin, 32*, 471–485.

36. http://www.slate.com/articles/health_and_science/science/2012/08/anti_gay_bigotry_online_analyzing_homophobic_comments_can_disarm_the_hate_.html.

37. Bass, L. (2007). *Out of sync: A memoir*. New York, NY: Simon Spotlight Entertainment.

38. For a thorough review, see Baumeister, R. F. (2000). Gender differences in erotic plasticity: The female sex drive as socially flexible and responsive. *Psychological Bulletin, 126*, 347–374; Diamond, L. M. (2008). Female bisexuality from adolescence to adulthood: Results from a 10-year longitudinal study. *Developmental Psychology, 44*, 5–14.

39. Whisman, V. (1996). *Queer by choice*. New York: Routledge; Savin-Williams, R. C. (1990). *Gay and lesbian youth: Expressions of identity*. New York: Hemisphere; Haslam, N., & Levy, S. R. (2007). Essentialist beliefs about homosexuality: Structure and implications for prejudice. *Personality and Social Psychology Bulletin, 32*, 471–485.

40. Inbar, Y., Pizarro, D. A., & Bloom, P. (2009). Conservatives are more easily disgusted than liberals. *Cognition and Emotion, 23*, 714–725.

41. Frias-Navarro, D., Monterde-I-Bort, H., Pascual-Soler, M., & Badenes-Ribera, L. (2015). Etiology of homosexuality and attitudes toward same-sex parenting. *Journal of Sex Research, 52*, 151–161. Also see Falomir-Pichastor, J. M., & Mugny, G. (2009). "I'm not gay. . . . I'm a real man!": Heterosexual men's gender self-esteem and sexual prejudice. *Personality and Social Psychology Bulletin, 35*(9), 1233–1243; Boysen, G. A., & Vogel, D. L. (2007). Biased assimilation and attitude polarization in response to learning about biological explanations of homosexuality. *Sex Roles, 57*, 755–762.

42. See p. 195 in Hamer, D., & Copeland, P. (1998). *Living with our genes*. New York, NY: Anchor Books.

43. For a review of the news coverage, see Conrad, P., & Markens, S. (2001). Constructing the "gay gene" in the news: Optimism and skepticism in the US and British press. *Health: An Interdisciplinary Journal for the Social Study of Health, Illness and Medicine, 5,* 373–400.

44. From the *Boston Globe,* July 20, 1993.

45. From Conrad, P., & Markens, S. (2001). Constructing the "gay gene" in the news: Optimism and skepticism in the US and British press. *Health: An Interdisciplinary Journal for the Social Study of Health, Illness and Medicine, 5,* 373–400.

46. http://www.pewforum.org/2012/07/31/two-thirds-of-democrats-now-support-gay-marriage-long-term-views-gay-marriage-adoption/.

47. Garretson, J., & Suhay, E. (2016). Scientific communication about biological influences on homosexuality and the politics of gay rights. *Political Research Quarterly, 69,* 17–29.

48. Quoted in Mayer, J. (2012, June 18). Bully pulpit. *New Yorker Magazine,* p. 58.

49. Iemmola, F., & Camperio Ciani, A. (2009). New evidence of genetic factors influencing sexual orientation in men: Female fecundity increase in the maternal line. *Archives of Sexual Behavior, 38,* 393–399.

50. King, M., Green, J., Osborn, D. P., Arkell, J., Hetherton, J., & Pereira, E. (2005). Family size in white gay and heterosexual men. *Archives of Sexual Behavior, 34,* 117–122; Yankelovich Partners (1994). *A Yankelovich Monitor perspective on gays/lesbians.* Chapel Hill, NC: Yankelovich Partners.

51. Lauber, J. (2009). *Chosen faith, chosen land: The untold story of America's 21st century Shakers.* Camden, ME: Down East Books.

52. For an excellent review of the Darwinian paradox, see LeVay, S. (2011). *Gay, straight, and the reason why.* Oxford, UK: Oxford University Press.

53. For a review, see Baumeister, R. F. (2000). Gender differences in erotic plasticity: The female sex drive as socially flexible and responsive. *Psychological Bulletin, 126,* 347–374.

54. Iemmola, F., & Camperio Ciani, A. (2009). New evidence of genetic factors influencing sexual orientation in men: Female fecundity increase in the maternal line. *Archives of Sexual Behavior, 38,* 393–399. Also see Camperio Ciani, A., Corna, F., & Capiluppi, C. (2004). Evidence for maternally inherited factors favouring male homosexuality and promoting female fecundity. *Proceedings of the Royal Society of London, Series G: Biological Sciences, 271,* 2217–2221; Rahman, Q., Collins, A., Morrison, M., Orrells, J. C., Cadinouche, K., Greenfield, S., et al. (2008). Maternal inheritance and familial fecundity factors in male homosexuality. *Archives of Sexual Behavior, 37,* 962–969.

55. Wilson, E. O. (1978). *On human nature*. Cambridge, MA: Harvard University Press.

56. Bobrow, D., & Bailey, J. M. (2001). Is male homosexuality maintained by kin selection? *Evolution and Human Behavior*, *22*, 361–368. Rahman, Q., & Hull, M. S. (2005). An empirical test of the kin selection hypothesis for male homosexuality. *Archives of Sexual Behavior*, *34*, 461–467.

57. Vasey, P. L., & VanderLaan, D. P. (2008). Avuncular tendencies and the evolution of male androphilia in Samoan Fa'afafine. *Archives of Sexual Behavior*, *40*, 495–503; Vasey, P. L., & VanderLaan, D. P. (2010). An adaptive cognitive disassociation between willingness to help kin and nonkin in Samoan Fa'afafine. *Psychological Science*, *21*, 292–297.

58. VanderLaan, D. P., Blanchard, R., Wood, H., & Zucker, K. J. (2014). Birth order and sibling sex ratio of children and adolescents referred to a gender identity service. *PLoS ONE 9*(3): e90257.

59. Blanchard, R., & Bogaert, A. F. (2004). Proportion of homosexual men who owe their sexual orientation to fraternal birth order: An estimate based on two national probability samples. *American Journal of Human Biology*, *16*, 151–157.

60. Bem, D. (1996). Exotic becomes erotic: A developmental theory of sexual orientation. *Psychological Review*, *103*, 320–335. Note that the differences between homosexual and heterosexual men are considerably larger than they are for women, and this account seems less applicable to lesbians. See Peplau, L. A., Garnets, L. D., Spalding, L. R., Conley, T. D., & Veniegas, R. C. (1998). A critique of Bem's "Exotic becomes erotic" theory of sexual orientation. *Psychological Review*, *105*, 387–394.

61. Bailey, J. M., & Zucker, K. J. (1995). Childhood sex-typed behavior and sexual orientation: A conceptual analysis and quantitative review. *Developmental Psychology*, *31*, 43–55.

62. Bell, A. P., Weinberg, M. S., & Hammersmith, S. K. (1981). *Sexual preference: Its development in men and women*. Bloomington, IN: Indiana Press; Carduso, F. L. (2009). Recalled sex-typed behavior in childhood and sports' preferences in adulthood of heterosexual, bisexual, and homosexual men from Brazil, Turkey, and Thailand. *Archives of Sexual Behavior*, *38*, 726–736; Green, R. (1987). *The "sissy boy syndrome" and the development of homosexuality*. New Haven, CT: Yale University Press.

63. Dutton, D. G., & Aron, A. P. (1974). Some evidence for heightened sexual attraction under conditions of high anxiety. *Journal of Personality and Social Psychology*, *30*, 510–517; White, G. L., Fishbein, S., & Rutstein, J. (1981). Passionate love and the misattribution of arousal. *Journal of Personality and Social Psychology*, *41*, 56–62.

64. See Bieber, I., Dain, H. J., Dince, P. R., Drellich, M. G., Grand, H. G., Gundlach, R. R., et al. (1962). *Homosexuality: A psychoanalytic study of male homosexuals.* New York, NY: Basic Books.

65. McKellen, I. (1993, July 22). Through a gay viewfinder. *Guardian.*

66. See Conrad, P., & Markens, S. (2001). Constructing the "gay gene" in the news: Optimism and skepticism in the US and British press. *Health: An Interdisciplinary Journal for the Social Study of Health, Illness and Medicine, 5,* 373–400.

67. Kitzinger, J. (2012). Constructing and deconstructing the "gay gene": Media reporting of genetics, sexual diversity, and "deviance." In R. Ellison & A. H. Goodman (Eds.), *The nature of difference: Science, society, and human biology* (pp. 99–118). Boca Raton, FL: Taylor and Francis.

68. Sheldon, J. P., Pfeffer, C. A., Jayaratne, T. E., Feldbaum, M., & Petty, E. M. (2007). Beliefs about the etiology of homosexuality and about the ramifications of discovering its possible genetic origin. *Journal of Homosexuality, 52,* 111–150.

69. Mustanski, B. S., Dupree, M. G., Nievergelt, C. M., Bocklandt, S., Schork, N. J., & Hamer, D. H. (2005). A genomewide scan of male sexual orientation. *Human Genetics, 116,* 272–278; Ramagopalan, S. V., Dyment, D. A., Handunnetthi, L., Rice, G. P., & Ebers, G. C. (2010). A genome-wide scan of male sexual orientation. *Journal of Human Genetics, 55,* 131–132; Sanders, A. R., Cao, Q., Zhang, J., Badner, J. A., Goldin, L. R., Guroff, J. J., et al. (1998). *Genetic linkage study of male homosexual orientation.* Toronto, ON: American Psychiatric Association; Rice, G., Anderson, C., Risch, N., & Ebers, G. (1999). Male homosexuality: Absence of linkage to microsatellite markers at Xq28. *Science, 284,* 665–667. But note that the largest study conducted to date purports to replicate Hamer's original findings, although still without specifying any individual genes: Sanders, A. R., Martin, E. R., Beecham, G. W., Guo, S., Dawood, K., Rieger, G., et al. (2014). Genome-wide scan demonstrates significant linkage for male sexual orientation. *Psychological Medicine, 45,* 1379–1388.

70. Rice, W. R., Friberg, U., & Gavrilets, S. (2012). Homosexuality as a consequence of epigenetically canalized sexual development. *The Quarterly Review of Biology, 87,* 343–368.

## 6. Race and Ancestry: How Our Genes Connect and Divide Us

1. From Puhl, J. (2014, April 3). Metamorphosis: A Hungarian extremist explores his Jewish roots. *Der Spiegel.* Retrieved from http://www.spiegel

.de/international/europe/a-hungarian-right-wing-extremist-explores-his
-jewish-roots-a-962156.html.

2. From Gorondi, P. (2014, June 10). Anti-Semitic, far-right politician's astonishing transformation after finding out he is a Jew. *National Post.* Retrieved from http://news.nationalpost.com/2014/06/10/anti-semitic -far-right-politicians-astonishing-transformation-after-finding-out -he-is-a-jew/.

3. From Applebaum, A. (2013, November 11). Anti-Semite and Jew. *The New Yorker.* Retrieved from http://www.newyorker.com/ magazine/2013/11/11/anti-semite-and-jew.

4. From Gorondi, P. (2014, June 10). Anti-Semitic, far-right politician's astonishing transformation after finding out he is a Jew. *National Post.* Retrieved from http://news.nationalpost.com/2014/06/10/ anti-semitic-far-right-politicians-astonishing-transformation -after-finding-out-he-is-a-jew/. Also see Ain, S. (2014, October 21). New life, New mission for Ex-Jobbik leader. *The Jewish Week.* Retrieved from http://www.thejewishweek.com/news/international/ new-life-new-mission-ex-jobbik-leader.

5. Owusu-Bempah, K. (2007). *Children and separation: Genealogical connectedness perspective.* East Sussex, UK: Psychology Press; Sants, H. J. (1964). Genealogical bewilderment in children with substitute parents. *British Journal of Medical Psychology, 37,* 133–141.

6. From Lifton, B. J. (1994). *Journey of the adopted self: A quest for wholeness* (p. 7). New York, NY: Basic Books.

7. From Lifton, B. J. (1979). Lost & found: The adoption experience (p. 8). New York, NY: Harper & Row.

8. For an extended exploration of our fascination with genealogy, see Zerubavel, E. (2012). *Ancestors & relatives. Genealogy, identity, & community.* New York, NY: Oxford University Press.

9. Rodriguez, G. (2014, May 30). How genealogy became almost as popular as porn. *Time Magazine.*

10. Royal, C. D., Novembre, J., Fullerton, S. M., Goldstein, D. B., Long, J. C., Bamshad, M. J., & Clark, A. G. (2010). Inferring genetic ancestry: Opportunities, challenges, and implications. *The American Journal of Human Genetics, 86,* 661–673.

11. Petrone, J. (2015, January 13). Consumer genomics market should pass "tipping point" of 3 million samples tested in 2015. Retrieved from https://www.genomeweb.com/microarrays-multiplexing/consumer -genomics-market-should-pass-tipping-point-3-million-samples-tested.

12. Calculating the rate of genetic mutations is key for estimating the precise dates of migration. See Scally, A., & Durbin, R. (2012). Revising the human mutation rate: Implications for understanding human evolution. *Nature Reviews Genetics, 13*, 745–753.

13. See Mellars, P., Gori, K. C., Carr, M., Soares, P. A., & Richards, M. B. (2013). Genetic and archaeological perspectives on the initial modern human colonization of southern Asia. *Proceedings of the National Academy of Sciences, 110*, 10699–10704.

14. For a detailed review of early human migrations, see Bellwood, P. (2014). *First migrants: Ancient migration in global perspective.* West Sussex, UK: Wiley.

15. There is also parallel evidence for the pattern of human migration by looking at the different language families that are spoken around the world. See Cavalli-Sforza, L. L. (2000). *Genes, people, and languages.* New York, NY: Farrar Straus & Giroux.

16. Ingman, M., Kaessmann, H., Paabo, S., & Gyllensten, U. (2000). Mitochondrial genome variation and the origin of modern humans. *Nature, 408*, 708–713.

17. Underhill, P. A., Shen, P., Lin, A. A., Jin, L., Passarino, G., Yang, W. H., et al. (2000). Y chromosome sequence variation and the history of human populations. *Nature Genetics, 26*, 358–361. More recent analyses estimate that the date for Y-chromosomal Adam might be as old as 156,000 years ago. Poznik, G. D., Henn, B. M., Yee, M. C., Sliwerska, E., Euskirchen. G. M., Lin, A. A., et al. (2013). Sequencing Y chromosomes resolves discrepancy in time to common ancestor of males versus females. *Science, 341*, 562–565.

18. The reason that our common paternal ancestor is more recent than our common maternal ancestor is that there is more variability among men than women in terms of how many offspring they produce. Some men have fathered hundreds of children, whereas most women typically have a fairly similar number of children, and this means that we don't have to go back as far to find our common paternal ancestor.

19. Bellis, C., Hughes, R. M., Begley, K. N., Quinlan, S., Lea, R. A., Heath, S. C., et al. (2005). Phenotypical characterisation of the isolated Norfolk Island population focusing on epidemiological indicators of cardiovascular disease. *Human Heredity, 60*, 211–219.

20. Chiao, J. Y., & Blizinsky, K. D. (2010). Culture–gene coevolution of individualism–collectivism and the serotonin transporter gene. *Proceedings of the Royal Society B, 277*, 529–537.

21. Novembre, J., Johnson, T., Bryc, K., Kutalik, Z., Boyko, A. R., Auton, A., et al. (2008). Genes mirror geography within Europe. *Nature*, *456*, 98–101.

22. See Elhaik, E., Tatarinova, T., Chebotarev, D., Piras, I. S., Maria Calò, C., De Montis, A., et al. (2014). Geographic population structure analysis of worldwide human populations infers their biogeographical origins. *Nature Communications*, *5*, 3513. doi: 10.1038/ncomms4513.

23. Estimating the actual percentage breakdown of each of your ancestors is somewhat limited because we inherit our DNA from our parents in chunks through meiosis. Because of this chunking, some of our distant ancestors do not leave any of their DNA in our genome. See Royal, C. D., Novembre, J., Fullerton, S. M., Goldstein, D. B., Long, J. C., Bamshad, M. J., et al. (2010). Inferring genetic ancestry: Opportunities, challenges, and implications. *The American Journal of Human Genetics*, *86*, 661–673.

24. For a detailed investigation of many African Americans' genetic testing experience, see Nelson, A. (2008). Bio science: Genetic ancestry testing and the pursuit of African ancestry. *Social Studies of Science*, *38*, 759–83.

25. Nelson, A. (2008). Genetic genealogy testing and the pursuit of African ancestry. *Social Studies of Science*, *38*, 759–783.

26. The quotes in this paragraph come from https://chancellorfiles.wordpress .com/2008/02/27/wayne-joseph-thought-he-was-black.

27. For world population data over time, see http://themasites.pbl.nl/tridion/ en/themasites/hyde/basicdrivingfactors/population/index-2.html.

28. For some more information on the rates of cousin-marriages around the world, see Bittles, A. H. (2001). Consanguinity and its relevance to clinical genetics. *Clinical Genetics*, *60*, 89–98.

29. See http://cnsnews.com/news/article/obama-distant-cousins-palin -limbaugh-bush and http://query.nytimes.com/gst/fullpage.html?res=99 00E4D61131F935A15750C0A96E9C8B63&ref=amyharmon.

30. Ralph, P., & Coop, G. (2013). The geography of recent genetic ancestry across Europe. *PLoS Biology*, *11*(5), e1001555. doi: 10.1371/journal. pbio.1001555.

31. Rohde, D. L. T., Olson, S., & Chang, J. T. (2004). Modeling the recent common ancestry of all living humans. *Nature*, *431*, 562–566.

32. Olson, S. (2002). *Mapping human history*. New York, NY: Houghton Mifflin.

33. For an argument about Jesus's children, see Baigent, M., Leigh, R., & Lincoln, H. (2004). *Holy blood, holy grail*. New York, NY: Dell.

34. See Olson, S. (2006, March 15). Why we're all Jesus's children. *Slate*.

Retrieved from http://www.slate.com/articles/health_and_science/science/2006/03/why_were_all_jesus_children.html.

35. Schutzenberger, A. A. (1998). *The ancestor syndrome: Transgenerational psychotherapy and the hidden links in the family tree.* London: Routledge.

36. Roth, W. D., & Lyon, K. (in press). Genetic ancestry tests and race: Who takes them, why, and how do they affect racial identities? In K. Suzuki & D. von Vacano (Eds.), *Reconsidering Race: Cross-Disciplinary and Interdisciplinary Approaches.* New York: Oxford University Press.

37. From http://www.raceandhistory.com/cgi-bin/forum/webbbs_config.pl?md=read;id=1400.

38. Yang, A. (2007). Is Oprah Zulu? Sampling and seeming certainty in DNA ancestry testing. *Chance, 20,* 32–39.

39. From http://www.slaverysite.com/Body/genealogy.htm.

40. The American Society of Human Genetics (2008, November 13). Ancestry testing statement. Retrieved from http://www.ashg.org/pdf/ashgancestrytestingstatement_final.pdf; Bolnick, D. A., Fullwiley, D., Duster, T., Cooper, R. S., Fujimura, J. H., Kahn, J., et al. (2007). The science and business of genetic ancestry testing. *Science, 318,* 399–400; Lee, S. S., Bolnick, D. A., Duster, T., Ossorio, P., & Tallbear, K. (2009). The illusive gold standard in genetic ancestry testing. *Science, 325,* 38–39; Lindee, S. (2013). Map your own genes! In S. Krimsky & J. Gruber (Eds.), *Genetic explanations: Sense and nonsense* (pp. 186–200). Cambridge, MA: Harvard University Press; Royal, C. D., Novembre, J., Fullerton, S. M., Goldstein, D. B., Long, J. C., Bamshad, M. J., & Clark, A. G. (2010). Inferring genetic ancestry: Opportunities, challenges, and implications. *The American Journal of Human Genetics, 86,* 661–673; Richards, M. (2003, February 14). Beware the gene genies, *Guardian.* Retrieved from http://www.guardian.co.uk/comment/story/0,,899835,00.html.

41. See Koerner, B. I. (2005, September). Blood feud. *Wired Magazine.* Retrieved from http://archive.wired.com/wired/archive/13.09/seminoles.html.

42. See Harmon, A. (2006, April 12). Seeking ancestry in DNA ties uncovered by tests. *New York Times.* Retrieved from http://www.nytimes.com/2006/04/12/us/12genes.html.

43. Travis, J. (2009, September 29). Scientists decry "flawed" and "horrifying" national tests. *Science Insider.* Retrieved from http://news.sciencemag.org/2009/09/scientists-decry-flawed-and-horrifying-nationality-tests; Hill, R., & Henderson, M. (2011, June 17). DNA test for bogus refu-

gees scrapped as expensive flop. *The Times*. Retrieved from http://www
.thetimes.co.uk/tto/news/uk/article3064981.ece.

44. Gil-White, F. J. (2001). Are ethnic groups biological "species" to the
human brain? Essentialism in our cognition of some social categories.
*Current Anthropology, 42,* 515–554; Rothbart, M., & Taylor, M. (1992).
Category labels and social reality: Do we view social categories as natural
kinds? In G. R. S. Semin & K. Fiedler (Eds.), *Language, interaction and
social cognition* (pp. 11–36). Newbury Park, CA: Sage.

45. Keller, J. (2005). In genes we trust: The biological component of psychological essentialism and its relationship to mechanisms of motivated social cognition. *Journal of Personality and Social Psychology,
88,* 686–702.

46. Allport, G. (1954). *The nature of prejudice* (p. 174). Reading, MA:
Addison-Wesley.

47. These items come from the Genetic Essentialist Tendencies measure.
Dar-Nimrod, I., Ruby, M. B., Cheung, B. Y., Tam, K., & Murray, D.
(2014, February). *The four horsemen of genetic essentialism: Theoretical
underpinnings, methodological advancements, and empirical findings.*
Symposium presented at the 2014 Society for Personality and Social
Psychology Annual Meeting, Austin, TX, USA.

48. These items come from the Social Dominance Orientation scale. Pratto,
F., Sidanius, J., Stallworth, L. M., & Malle, B. F. (1994). Social dominance orientation: A personality variable predicting social and political
attitudes. *Journal of Personality and Social Psychology, 67,* 741–763.

49. These items are from the Modern Racism Scale. McConahay, J. B.
(1986). Modern racism, ambivalence, and the Modern Racism Scale.
In J. F. Dovidio & S. L. Gaertner (Eds.), *Prejudice, discrimination, and
racism* (pp. 91–125). San Diego, CA: Academic Press.

50. These items are from the Right-wing Authoritarianism scale. Rattazzi,
A. M. M., & Canova, A. B. L. (2007). A short version of the Right-Wing
Authoritarianism (RWA) Scale. *Personality and Individual Differences,
43,* 1223–1234.

51. Across a few different studies, the correlation between genetic essentialist
tendencies (GET) and social dominance was $r = .24$, $r = .25$, and $r = .35$;
GET and right wing authoritarianism was $r = .21$ and $r = .30$; and GET
and modern racism was $r = .27$ and $r = .24$. Cheung, B. Y., Ream, A., &
Heine, S. J. (2015). *Correlations of the Genetic Essentialist Tendencies Scale.*
Unpublished.

52. See studies in Keller, J. (2005). In genes we trust: The biological component of psychological essentialism and its relationship to mechanisms of

motivated social cognition. *Journal of Personality and Social Psychology*, *88*, 686–702; Jayaratne, T. E., Ybarra, O., Sheldon, J. P., Brown, T. N., Feldbaum, M., Pfeffer, C. A., & Petty, E. M. (2006). White Americans' genetic lay theories of race differences and sexual orientation: Their relationship with prejudice toward Blacks, gay men and lesbians. *Group Processes & Intergroup Relations*, *9*, 77–94; Jayaratne, T., Gelman, S., Feldbaum, M., Sheldon, J., Petty, E., & Kardia, S. (2009). The perennial debate: Nature, nurture, or choice? Black and White Americans' explanations for individual differences. *Review of General Psychology*, *13*, 24–33.

53. http://www.nbcnews.com/id/21362732/ns/us_news-life/t/race-remarks -get-nobel-winner-trouble/#.VVZ30Dcqbd4.

54. http://indiancountrytodaymedianetwork.com/2007/12/21/lyons-curious -return-race-2007-91944.

55. See Sampanis, M. (2003). *Preserving power through coalitions: Comparing the grand strategy of Great Britain and the United States*. Santa Barbara, CA: Praeger.

56. The letter denouncing Wade's book was retrieved from http://cehg .stanford.edu/letter-from-population-geneticists/.

57. I'd recommend my own book on this topic: Heine, S. J. (2016). *Cultural Psychology* (3rd ed.). New York, NY: W. W. Norton. Also see Diamond, J. (1998). *Guns, germs, and steel*. New York, NY: W. W. Norton; Acemoglu, D., & Robinson, J. A. (2013). *Why nations fail: The origins of power, prosperity, and poverty*. New York, NY: Crown; Landes, D. S. (1999). *The wealth and poverty of nations*. New York, NY: W. W. Norton; Ferguson, N. (2011). *Civilization: The West and the rest*. New York, NY: Penguin.

58. Haidt, J. (2012). *The righteous mind* (p. 429). New York, NY: Vintage Books.

59. See Needham, J. (1956). *Science and civilisation in China, Vol. 2. History of scientific thought*. Cambridge, UK: Cambridge University Press.

60. For a detailed discussion of the controversy and faulty science in the first race-based medicine, see Kahn, J. (2014). *Race in a bottle: The story of Bidil and racialized medicine in a post-genomic age*. New York, NY: Columbia University Press; Bliss, C. (2012). *Race decoded: The genomic fight for social justice*. Stanford, CA: Stanford University Press; Jones, D. (2013). The prospects of personalized medicine. In S. Krimsky & J. Gruber (Eds.), *Genetic explanations: Sense and Nonsense*. Cambridge, MA: Harvard University Press.

61. Center for Disease Control (2004). *National Center for Health Statistics*.

Retrieved from http://www.cdc.gov/nchs/data/hus/tables/2003/03hus 066.pdf.

62. Cooper, R. S., Rotimi, C., Ataman, S., McGee, D., Osotimehin, B., Kadiri, S., et al. (1997). The prevalence of hypertension in seven populations of West African origin. *American Journal of Public Health*, *87*, 160–68.

    Akinkugbe, O. O. (1987). World epidemiology of hypertension in Blacks. *Journal of Clinical Hypertension*, *3*, 1s–8s.

63. Wilson, T. W., & Grim, C. E. (1991). Biohistory of slavery and blood pressure differences in blacks today. A hypothesis. *Hypertension*, *17*, I122–I129.

    Grim, C. E., & Robinson M. (1996). Blood pressure variation in blacks: Genetic factors. *Seminars in Nephrology*, *16*, 83–93.

64. For a damning critique of each aspect of the slavery hypothesis, see Curtin, P. D. (1992). The slavery hypothesis for hypertension among African Americans: The historical evidence. *American Journal of Public Health*, *82*, 1681–86. Also see Cooper, R., & Rotimi, C. (1994). Hypertension in populations of West African origin: Is there a genetic predisposition? *Journal of Hypertension*, *12*, 215–227; Dressler, W. W., Oths, K. S., & Gravlee, C. C. (2005). Race and ethnicity in public health research: Models to explain health disparities. *Annual Review of Anthropology*, *34*, 231–252.

65. Kaufman, J. S., & Hall, S. A. (2003). The slavery hypertension hypothesis: Dissemination and appeal of a modern race theory. *Epidemiology*, *14*, 111–118.

66. Also see Gravlee, C. C., Dressler, W. W., & Bernard, H. R. (2005). Skin color, social classification, and blood pressure in Southeastern Puerto Rico. *American Journal of Public Health*, *95*, 2191–2197; Sweet, E., McDade, T. W., Kiefe, C. I., & Liu, K. (2007). Relationships between skin color, income, and blood pressure among African Americans in the CARDIA study. *American Journal of Public Health*, *97*, 2253–2259; Non, A. L., Gravlee, C. C., & Mulligan, C. J. (2012). Education, genetic ancestry, and blood pressure in African Americans and Whites. *American Journal of Public Health*, *102*, 1559–1565.

67. Cavalli-Sforza, L. L., & Feldman, M. W. (2003). The application of molecular genetic approaches to the study of human evolution. *Nature Genetics*, *33*, 266–275; Caspari, R. (2010). Deconstructing race: Racial thinking, geographic variation, and implications for biological anthropology. In C. S. Larsen (Ed.), *A companion to biological anthropology* (pp. 104–122). Malden, MA: Wiley-Blackwell; Ramachandran, S.,

Deshpande, O., Roseman, C. C., Rosenberg, N. A., Feldman, M. W., & Cavalli-Sforza, L. L. (2005). Support from the relationship of genetic and geographic distance in human populations for a serial founder effect originating in Africa. *Proceedings of the National Academy of Sciences, 102,* 15942–15947.

68. For a thorough review of this argument, see Templeton, A. R. (2013). Biological races in humans. *Studies in History and Philosophy of Science Part C: Studies in History and Philosophy of Biological and Biomedical Sciences, 44,* 262–271.

69. See Smith, H. M., Chiszar, D., & Montanucci, R. R. (1997). Subspecies and classification. *Herpetological Review, 28,* 13–16.

70. Weiss, K. M., & Long, J. C. (2009). Non-Darwinian estimation: My ancestors, my genes' ancestors. *Genome Research, 19,* 703–710.

    Relethford, J. H. (2009). Race and global patterns of phenotypic variation. *American Journal of Physical Anthropology, 139,* 16–22.

71. See Smith, H. M., Chiszar, D., & Montanucci, R. R. (1997). Subspecies and classification. *Herpetological Review, 28,* 13–16.

72. For a review, see Templeton, A. R. (2013). Biological races in humans. *Studies in History and Philosophy of Science Part C: Studies in History and Philosophy of Biological and Biomedical Sciences, 44,* 262–271. The 4.3 percent estimate comes from Rosenberg, N. A., Pritchard, J. K., Weber, J. L., Cann, H. M., Kidd, K. K., Zhivotovsky, L. A., et al. (2002). Genetic structure of human populations. *Science, 298,* 2381–2385. Using a much smaller collection of genetic markers, Richard Lewontin had earlier made famous the notion that human races accounted for 6.3 percent of human genetic variability. Lewontin, R. (1972). The apportionment of human diversity. *Evolutionary Biology, 6,* 396–397.

73. Hunley, K. L., Healy, M. E., & Long, J. C. (2009). The global pattern of gene identity variation reveals a history of long-range migrations, bottlenecks, and local mate exchange: Implications for biological race. *American Journal of Physical Anthropology, 139,* 35–46; Templeton, A. R. (1998). Human races: A genetic and evolutionary perspective. *American Anthropologist, 100,* 632–650.

74. For a nice discussion of this point, see Henrich, J. (2015). *The secret of our success: How culture is driving human evolution, domesticating our species, and making us smarter.* Princeton, NJ: Princeton University Press.

75. A 2002 study by Rosenberg, N. A., Pritchard, J. K., Weber, J. L., Cann, H. M., Kidd, K. K., Zhivotovsky, L. A., et al. (Genetic structure of human populations. *Science, 298,* 2381–2385) is often cited in support of the idea that people's genes cluster to reveal five continental races (African,

European, Asian, Australian, and Native American; for example, see a lengthy discussion about this in Nicholas Wade's 2014 book *A troublesome inheritance* (New York, NY: Penguin Press). However, there are two points to note about this study—first, the method that is used is not exploring clustering of genes that have been selected through natural selection, but is largely based on genes that vary due to genetic drift. That is, it largely speaks to which groups of people have been interbreeding the most with each other rather than there being clusters of adaptive traits. Second, and more problematic for Wade's interpretation, is that this study does not provide support for five different continental clusters. The number of clusters that emerge in this analysis is determined by the researcher, and this study found support for any number of clusters that ranged between 4 and 20. There is no good evidence to suggest that human genes cluster into five continental groups that correspond to laypeople's thoughts about race. See Bolnick, D. A. (2008). Individual ancestry inference and the reification of race as a biological phenomenon. In Koenig, B. A., Lee, S. S., & Richardson, S. S. (Eds.), *Revisiting Race in a Genomic Age* (pp. 70–101). New Brunswick, NJ: Rutgers University Press; Rosenberg, N. A., Pritchard, J. K., Weber, J. L., Cann, H. M., Kidd, K. K., Zhivotovsky, L. A., et al. (2002). Genetic structure of human populations. *Science, 298*, 2381–2385.

76. Jorde, L. B., & Wooding, S. P. (2004). Genetic variation, classification and "race." *Nature Genetics, 36*, S28–S33; Gravlee, C. G. (2009). How race becomes biology: Embodiment of social inequality. *American Journal of Physical Anthropology, 139*, 47–57.

77. Coop, G., Pickrell, J. K., Novembre, J., Kudaravalli, S., Li, J., Absher, D., et al. (2009). The role of geography in human adaptation. *PLoS Genetics, 5*, e1000500.

78. Wade, N. (2014). *A troublesome inheritance.* New York, NY: Penguin Press.

79. Also see Yudell, M., Roberts, D., DeSalle, R., & Tishkoff, S. (2016). Taking race out of human genetics. *Science, 351*, 564–565.

80. Templeton, A. R. (2013). Biological races in humans. *Studies in History and Philosophy of Science Part C: Studies in History and Philosophy of Biological and Biomedical Sciences, 44*, 262–271. Also see Santos, R. V., Fry, P. H., Monteiro, S., Maio, M. C., Rodrigues, J. C., Bastos-Rodrigues, L., et al. (2009). Color, race, and genomic ancestry in Brazil. *Current Anthropology, 50*, 787–819.

81. But note that the degree to which hypodescent is formalized in practice varies a lot across groups. Many Native American tribes require a substantial fraction of Native ancestry to qualify for tribal membership.

For a discussion, see Snipp, C. M. (2003). Racial measurement in the American census: Past practices and implications for the future. *Annual Review of Sociology, 29*, 563–588.

82. Grant, M. (1916). *The passing of the great race.* New York, NY: Scribner.

83. Ho, A. K., Sidanius, J., Levin, D. T., & Banaji, M. R. (2011). Evidence for hypodescent and racial hierarchy in the categorization and perception of biracial individuals. *Journal of Personality and Social Psychology, 100*, 492–506.

84. Williams, M., & Eberhardt, J. (2008). Biological conceptions of race and the motivation to cross racial boundaries. *Journal of Personality and Social Psychology, 94*, 1033–1047.

85. Also see Phelan, J. C., Link, S. Z., & Yang, L. H. (2014). Direct-to-consumer racial admixture tests and beliefs about essential racial differences. *Social Psychology Quarterly, 77*, 296–318; and Phelan, J. C., Link, B. G., & Feldman, N. M. (2013). The genomic revolution and beliefs about essential racial differences: A backdoor to eugenics? *American Sociological Review, 78*, 167–191.

86. No, S., Hong, Y., Liao, H., Lee, K., Wood, D., & Chao, M. (2008). Lay theory of race affects and moderates Asian Americans' responses toward American culture. *Journal of Personality and Social Psychology, 95*, 991–1004. For conceptually similar findings, see Chao, M., Chen, J., Roisman, G., & Hong, Y. (2007). Essentializing race: Implications for bicultural individuals' cognition and physiological reactivity. *Psychological Science, 18*, 341–348.

87. Ambrose, S. H. (1998). Late Pleistocene human population bottlenecks, volcanic winter, and differentiation of modern humans. *Journal of Human Evolution, 34*, 623–651.

88. Fu, Q., Mittnik, A., Johnson, P. L., Bos, K., Lari, M., Bollongino, R., et al. (2013). A revised timescale for human evolution based on ancient mitochondrial genomes. *Current Biology, 23*, 553–559.

89. Gonder, M. K., Locatelli, S., Ghobrial, L., Mitchell, M. W., Kujawski, J. T., Lankester, F. J., et al. (2011). Evidence from Cameroon reveals differences in the genetic structure and histories of chimpanzee populations. *Proceedings of the National Academy of Science, 108*, 4766–4771.

90. Becquet, C., Patterson, N., Stone, A. C., Przeworski, M., & Reich, D. (2007). Genetic structure of chimpanzee populations. *PLoS Genetics, 3*, e66. doi: 10.1371/ journal.pgen.0030066.

91. http://www.cnn.com/TRANSCRIPTS/0006/26/bn.01.html

92. As quoted in Collins, F. S. (2010). *The language of life* (p. 147). New York, NY: Harper.

93. Kimel, S. Y., Huesmann, R., Kunst, J. R., & Halperin, E. (2016). Living in a genetic world: How learning about interethnic genetic similarities and differences affects peace and conflict. *Personality and Social Psychology Bulletin, 42*, 688–700.

94. Schmalor, A., Cheung, B. Y., & Heine, S. J. (2015). Exploring people's thoughts about the causes of ethnic stereotypes. Unpublished data. University of British Columbia.

95. For another example of a benefit of Story Two, see Kang, S. K., Plaks, J. E., & Remedios, J. D. (2015). Folk beliefs about genetic variation predict avoidance of biracial individuals. *Frontiers in Psychology, 6*, 357. Also see Plaks, J. E., Malahy, L. W., Sedlins, M., & Shoda, Y. (2012). Folk beliefs about human genetic variations predict discrete versus continuous racial categorization and evaluative bias. *Social Psychological and Personality Science, 3*, 31–39.

## 7. Essences and the Seductive Allure of Eugenics

1. Burt, C. (1934). Studying the minds of others. In C. Burt (Ed.), *How the mind works* (p. 28). London, England: Unwin Brothers.

2. Deary, I. J., Lawn, M., & Bartholomew, D. J. (2008). A conversation between Charles Spearman, Godfrey Thomson, and Edward L. Thorndike: The International Examinations Inquiry Meetings 1931–1938 (Vol. 11, p. 122). *History of Psychology, 11*, 163.

3. Heine, S. J., Kitayama, S., Lehman, D. R., Takata, T., Ide, E., Leung, C., & Matsumoto, H. (2001). Divergent consequences of success and failure in Japan and North America: An investigation of self-improving motivations and malleable selves. *Journal of Personality and Social Psychology, 81*, 599–615.

4. The correct answer is "square": you can have a square deal, a square meal, or you can be a square peg.

5. For a very detailed and engaging exploration of this work, see Dweck, C. (2007). *Mindset: The new psychology of success*. New York, NY: Ballantine.

6. Asbury, K., & Plomin, R. (2014). *G is for genes*. West Sussex, UK: Wiley.

7. Nisbett, R. E. (2009). *Intelligence and how to get it* (pp. 45–46). New York, NY: W. W. Norton.

8. Chabris, C. F., Hebert, B. M., Benjamin, D. J., Beauchamp, J. P., Cesarini, D., van der Loos, M. J. H. M., et al. (2012). Most reported genetic associations with general intelligence are probably false positives. *Psychological Science, 23*, 1314–1323.

9. Kirkpatrick, R. M., McGue, M., Iacono, W. G., Miller, M. B., & Basu, S. (2014). Results of a "GWAS Plus": General cognitive ability is substantially heritable and massively polygenic. *PLoS One, 9*, e112390; Butcher, L. M., Davis, O. S. P., Craig, I. W., & Plomin, R. (2008). Genome-wide quantitative trait locus association scan of general cognitive ability using pooled DNA and 500K single nucleotide polymorphism microarrays. *Genes, Brain and Behavior, 7*, 435–446; Franic, S., Dolan, C. V., Broxholme, J., Hu, H., Zemojtel, J., Davies, G. E., et al. (2015). Mendelian and polygenic inheritance of intelligence: A common set of causal genes. Using next-generation sequencing to examine the effects of 168 intellectual disability genes on normal-range intelligence. *Intelligence, 49*, 10–22.

10. Rietveld, C. A., Esko, T., Davies, G., Pers, T. H., Turley, P., Benyamin, B., et al. (2014). Common genetic variants associated with cognitive performance identified using the proxy-phenotype method. *Proceedings of the National Academy of Sciences, 111*, 13790–13794.

11. Davies, G., Tenesa, A., Payton, A., Yang, J., Harris, S. E., Liewald, D., et al. (2011). Genome-wide association studies establish that human intelligence is highly heritable and polygenic. *Molecular Psychiatry, 16*, 996–1005; Chabris, C. F., Hebert, B. M., Benjamin, D. J., Beauchamp, J. P., Cesarini, D., van der Loos, M. J. H. M., et al. (2012). Most reported genetic associations with general intelligence are probably false positives. *Psychological Science, 23*, 1314–1323.

12. Johnson, W. (2010). Understanding the genetics of intelligence: Can height help? Can corn oil? *Current Directions in Psychological Science, 19*, 177–182; Benyamin, B., Pourcain, B., Davis, O. S., Davies, G., Hansell, N. K., Brion, M. J., et al. (2014). Childhood intelligence is heritable, highly polygenic and associated with FNBP1L. *Molecular Psychiatry, 19*, 253–258.

13. A recent study found that a genome-wide analysis of educational attainment allowed researchers to predict approximately 2.5 IQ points for every standard deviation of individual's overall polygenic scores as calculated across the SNP chip. See Belsky, D. W., Moffitt, T. E., Corcoran, D. L., Domingue, B., Harrington, H., Hogan, S., et al. (in press). The genetics of success: How single-nucleotide polymorphisms associated with educational attainment relate to life-course development. *Psychological Science*.

14. Bouchard, T. J., Jr. (1998). Genetic and environmental influences on adult intelligence and special mental abilities. *Human Biology, 70*, 257–279; Deary, I. J., Johnson, W., & Houlihan, L. (2009). Genetic foundations of human intelligence. *Human Genetics, 126*, 215–232.

15. For an extended argument on this, see Nisbett, R. E., Aronson, J., Blair, C., Dickens, W., Flynn, J., Halpern, D. F., et al. (2012). Intelligence: New findings and theoretical developments. *American Psychologist, 67*, 130–159.

16. Scarr-Salapatek, S. (1971). Race, social class, and IQ. *Science, 174*, 1285–1295; Rowe, D. C., Jacobson, K. C., & Van den Oord, E. J. C. G. (1999). Genetic and environmental influences on vocabulary IQ: Parental education level as moderator. *Child Development, 70*, 1151–1162; Turkheimer, E., Haley, A., Waldron, M., D'Onofrio, B., & Gottesman, I. (2003). Socioeconomic status modifies heritability of IQ in young children. *Psychological Science, 14*, 623–628. Curiously, the effects of social class on heritability of IQ appear more pronounced in the United States than elsewhere (see Tucker-Drop, E. M., & Bates, T. C. [in press]. Large cross-national differences in gene x socioeconomic status interaction on intelligence. *Psychological Science*). This might be a product of larger income inequality. In places with a more developed social welfare state, such as in Europe, there aren't such reliable differences in the estimates of heritability for IQ between the wealthy and the poor.

17. For more discussion of this, see Nisbett, R. E. (2009). *Intelligence and how to get it.* New York, NY: W. W. Norton; and Stoolmiller, M. (1999). Implications of the restricted range of family environments for estimates of heritability and nonshared environment in behavior-genetic adoption studies. *Psychological Bulletin, 125*, 392–409.

18. Nisbett, R. E. (2009). *Intelligence and how to get it.* New York, NY: W. W. Norton.

19. Stoolmiller, M. (1999). Implications of the restricted range of family environments for estimates of heritability and nonshared environment in behavior-genetic adoption studies. *Psychological Bulletin, 125*, 392–409.

20. Dillman, D. A. (1978). *Mail and telephone surveys: The Total Design Method.* New York, NY: Wiley. See the discussion in Nisbett, R. E., Aronson, J., Blair, C., Dickens, W., Flynn, J., Halpern, D. F., et al. (2012). Intelligence: New findings and theoretical developments. *American Psychologist, 67*, 130–159.

21. For a detailed review of this point, see Henrich, J., Heine, S. J., & Norenzayan, A. (2010). The weirdest people in the world. *Behavioral and Brain Sciences, 33*, 61–83.

22. See a nice discussion of this point on pp. 37–38 in Nisbett, R. E. (2009). *Intelligence and how to get it.* New York, NY: W. W. Norton.

23. Greulich, W. W. (1957). A comparison of the physical growth and

development of American-born and native Japanese children. *American Journal of Physical Anthropology, 15,* 489–515.

24. Nisbett, R. E., Aronson, J., Blair, C., Dickens, W., Flynn, J., Halpern, D. F., et al. (2012). Intelligence: New findings and theoretical developments. *American Psychologist, 67,* 130–159; Duyme, M., Dumaret, A., & Tomkiewicz, S. (1999). How can we boost IQs of "dull" children? A late adoption study. *Proceedings of the National Academy of Sciences, USA, 96,* 8790–8794; Moore, E. C. J. (1986). Family socialization and the IQ test performance of traditionally and trans-racially adopted Black children. *Developmental Psychology, 22,* 317–326; Van IJzendoorn, M. H., Juffer, F., & Poelhuis, C. W. K. (2005). Adoption and cognitive development: A meta-analytic comparison of adopted and nonadopted children's IQ and school performance. *Psychological Bulletin, 131,* 301–316.

25. Brockman, J. (Ed). (2006). *What is your dangerous idea?* New York, NY: Harper.

26. Herrnstein, R. J., & Murray, C. (1994). *The bell curve.* New York, NY: Free Press.

27. Snyderman, M., & Rothman, S. (1988). *The IQ controversy: The media and public policy.* New Brunswick, NJ: Transaction Books.

28. Rosenberg, N. A., Pritchard, J. K., Weber, J. L., Cann, H. M., Kidd, K. K., Zhivotovsky, L. A., et al. (2002). Genetic structure of human populations. *Science, 298,* 2381–2385.

29. Dickens, W. T., & Flynn, J. R. (2006). Black Americans reduce the racial IQ gap: Evidence from standardization samples. *Psychological Science, 17,* 913–920.

30. Nisbett, R. E., Aronson, J., Blair, C., Dickens, W., Flynn, J., Halpern, D. F., et al. (2012). Intelligence: New findings and theoretical developments. *American Psychologist, 67,* 130–159.

31. Ogbu, J. U. (1994). *Minority education and caste. The American system in cross-cultural perspective.* New York, NY: Academic Press.

32. You can also see evidence of the different kinds of educational opportunities provided by families across the races. Black children who were adopted by White families had IQ scores that were about 13 points higher than those Black children adopted by Black families. Of note, those adoptees who had two Black biological parents did not differ in their IQ from those adoptees who had one Black and one White biological parent, which is yet another reason to reject racial theories for differences in IQ. See Moore, E. G. J. (1986). Family socialization and

the IQ test performance of traditionally and trans-racially adopted Black children. *Developmental Psychology, 22,* 317–326.

33. Nisbett, R. E., Aronson, J., Blair, C., Dickens, W., Flynn, J., Halpern, D. F., et al. (2012). Intelligence: New findings and theoretical developments. *American Psychologist, 67,* 130–159; Duyme, M., Dumaret, A., & Tomkiewicz, S. (1999). How can we boost IQs of "dull" children? A late adoption study. *Proceedings of the National Academy of Sciences, USA, 96,* 8790–8794; Moore, E. C. J. (1986). Family socialization and the IQ test performance of traditionally and trans-racially adopted Black children. *Developmental Psychology, 22,* 317–326; Van Ijzendoorn, M. H., Juffer, F., & Poelhuis, C. W. K. (2005). Adoption and cognitive development: A meta-analytic comparison of adopted and nonadopted children's IQ and school performance. *Psychological Bulletin, 131,* 301–316.

34. See Flynn, J. R. (2007). *What is intelligence?* Cambridge, UK: Cambridge University Press.

35. Pietschnig, J., & Voracek, M. (2015). One century of global IQ gains: A formal meta-analysis of the Flynn Effect (1909–2013). *Perspectives on Psychological Science, 10,* 282–306.

36. Trahan, L., Stuebing, K. K., Hiscock, M. K., & Fletcher, J. M. (2014). The Flynn Effect: A Meta-analysis. *Psychological Bulletin, 140,* 1332–1360. However, some analyses suggest that the slope of increases has flattened slightly more recently: Pietschnig, J., & Voracek, M. (2015). One century of global IQ gains: A formal meta-analysis of the Flynn Effect (1909–2013). *Perspectives on Psychological Science, 10,* 282–306.

37. For a discussion, see Nisbett, R. E., Aronson, J., Blair, C., Dickens, W., Flynn, J., Halpern, D. F., et al. (2012). Intelligence: New findings and theoretical developments. *American Psychologist, 67,* 130–159.

38. The correct answer to this Raven's item is "2."

39. Healy, J. M. (1990). *Endangered minds: Why children don't think and what we can do about it.* New York: Simon & Schuster.

40. Goldin, C. (1998). America's graduation from high school: The evolution and spread of secondary schooling in the twentieth century. *Journal of Economic History, 58,* 345–374; US Census Bureau. (2014). Selected social characteristics in the United States: 2012 American community survey 1-year estimates. Retrieved from factfinder.census.gov/faces/tableservices/jsf/pages/productview .xhtml?pid=ACS_12_1YR_DP02&prodType=table.

41. Johnson, S. (2005). *Everything bad is good for you.* New York, NY: Riverhead Books.

42. From the World War 1 Alpha Test, which was used to test the intel-

ligence of literate army recruits. See p. 136 in Rafter, N. (2008). *The criminal mind*. New York, NY: New York University Press. Also see p. 66 in Paul, D. B. (1995). *Controlling human heredity: 1865 to the present*. Amherst, NY: Humanity Books.

43. Raven, J. C. (1965). *Guide to using the coloured progressive matrices*. London: H. K. Lewis.

44. Davis, H. E. (2014). *Variable education exposure and cognitive task performance among the Tsimane forager-horticulturalists*. Doctoral dissertation. The University of New Mexico.

45. Gordon, P. (2005). Numerical cognition without words: Evidence from Amazonia. *Science, 306*, 496–499. Also see Everett, D. L. (2005). Cultural constraints on grammar and cognition in Pirahã. *Current Anthropology, 46*, 621–634.

46. These data are not yet published. They are part of research conducted by Duncan Stibbard-Hawkes and Coren Apicella. Personal communication from Coren Apicella, June 9, 2015.

47. Raichlen, D. A., Wood, B. M., Gordon, A. D., Mabulla, A. Z. P., Marlowe, F. W., & Pontzer, H. (2014). Evidence of Levy walk foraging patterns in human hunter-gatherers. *Proceedings of the National Academy of Sciences, 111*, 728–733.

48. See p. 640 of Anderson, M. (1971). Bad seed. In S. Richards (Ed.), *Best mystery and suspense plays of the modern theatre* (pp. 597–667). New York, NY: Dodd, Mead.

49. Maudsley, H. (1874). *Responsibility in mental disease* (pp. 25, 33). London: Henry S. King.

50. Rafter, N. (2008). *The criminal brain* (p. 65). New York, NY: New York University Press.

51. See p. 51 of Lombroso, C. (2006). *Criminal man* (M. Gibson & N. H. Rafter, Trans.). Durham, NC: Duke University Press.

52. Horn, D. (2003). *The criminal body: Lombroso and the anatomy of deviance*. New York, NY: Routledge.

53. However, a facial width-to-height ratio has been found to predict which hockey players will be most aggressive. Carre, J. M., & McCormick, C. M. (2008). In your face: facial metrics predict aggressive behavior in the laboratory and in varsity and professional hockey players. *Proceedings of the Royal Society B, 275*, 2651–2656.

54. Brunner, H. G., Nelen, M. R., van Zandvoort, P., Abeling, N. G., van Gennip, A. H., Wolters, E. C., et al. (1993). X-linked borderline mental retardation with prominent behavioral disturbance: Phenotype, genetic localization, and evidence for disturbed monoamine metabolism.

*American Journal of Human Genetics, 52,* 1032–1039; Brunner, H. G., Nelen, M., Breakefield, X. O., Ropers, H. H., & van Oost, B. A. (1993). Abnormal behavior associated with a point mutation in the structural gene for monoamine oxidase A. *Science, 262,* 578–580.

55. Caspi, A., McClay, J., Moffitt, T. E., Mill, J., Martin, J., Craig, I. W., et al. (2002). Role of genotype in the cycle of violence in maltreated children. *Science, 297,* 851–854. For a recent meta-analysis documenting the robustness of this finding, see Byrd, A. L., & Manuck, S. B. (2014). MAOA, childhood maltreatment and antisocial behavior: Meta-analysis of a gene-environment interaction. *Biological Psychiatry, 75,* 9–17.

56. Lea, R., & Chambers, G. (2007). Monoamine oxidase, addiction, and the "warrior" gene hypothesis. *New Zealand Medical Journal, 120* (1250), 5–10; Beaver, K. M., Davis, C., Leavitt, K., Schriger, I., Benson, K., Bhakta, S., et al. (2012). Exploring the association between the 2-repeat allele of the MAOA gene promoter polymorphism and psychopathic personality traits, arrests, incarceration, and lifetime antisocial behavior. *Personality and Individual Differences, 54,* 164–168.

57. Wade, N. (2014). *A troublesome inheritance.* New York, NY: Penguin Press. Stokes, J. (2007, March 5). Scientist defends "warrior" gene. *The New Zealand Herald.*

58. Widon, C. S., & Brzustowicz, L. M. (2006). MAOA and the "cycle of violence": Childhood abuse and neglect, MAOA genotype, and risk for violent and antisocial behavior. *Biological Psychiatry, 60,* 684–689.

59. For a list of these and many other associates of the MAOA gene, see Charney, E., & English, W. (2012). Candidate genes and political behavior. *American Political Science Review, 106,* 1–34.

60. Global Study on Homicide 2013: Trends, contexts, data. Retrieved from http://www.unodc.org/documents/gsh/pdfs/2014_GLOBAL_HOMICIDE_BOOK_web.pdf.

61. Way, B. M., & Lieberman, M. D. (2010). Is there a genetic contribution to cultural differences? Collectivism, individualism and genetic markers of social sensitivity. *Social and Cognitive Affective Neuroscience, 5,* 203–211; Lea, R., & Chambers, G. (2007). Monoamine oxidase, addiction, and the "warrior" gene hypothesis. *New Zealand Medical Journal, 120*(1250), 5–10.

62. http://www.nature.com/news/2009/091030/full/news.2009.1050.html

63. Aspinwall, L. G., Brown, T. R., & Tabery, J. (2012). The double-edged sword: Does biomechanism increase or decrease judges' sentencing of psychopaths? *Science, 337*(6096), 846–849.

64. Cheung, B. Y., & Heine, S. J. (in press). The double-edged sword of

genetic accounts of criminality: Causal attributions from genetic ascriptions affect legal decision-making. *Personality and Social Psychology Bulletin.*

65. For another example of this effect, see Monterosso, J., Royzman, E. B., & Schwartz, B. (2005). Explaining away responsibility: Effects of scientific explanation on perceived culpability. *Ethics & Behavior, 15,* 139–158.

66. Daly, M., & Wilson, M. (1988). *Homicide.* New York, NY: Aldine de Gruyter.

67. Morse, S. J. (2006). Brain overclaim syndrome and criminal responsibility: A diagnostic note. *Ohio State Journal of Criminal Law, 3,* 397–412.

68. United States Court of Appeals, Second Circuit. United States of America, Appellee, v. Gary Cossey, Defendant-Appellant. Docket No. 09-5170-CR. Retrieved from http://caselaw.findlaw.com/us-2nd-circuit/1554381.html.

69. Kenneally, C. (2014). *The invisible history of the human race.* New York, NY: Viking.

70. Kevles, D. J. (1995). *In the name of eugenics.* Cambridge, MA: Harvard University Press; Paul, D. B. (1995). *Controlling human heredity: 1865 to the present.* New York, NY: Humanity Books; Kevles, D. J. (1992). Out of eugenics: The historical politics of the human genome. In D. J. Kevles & L. Hood (Eds.) *The code of codes* (pp. 3–36). Cambridge, MA: Harvard University Press.

71. Lombardo, P. A. (2008). *Three generations, no imbeciles.* (p. 117). Baltimore, MD: Johns Hopkins University Press.

72. Leuchtenburg, W. E. (1995). *The Supreme Court reborn.* New York, NY: Oxford University Press.

73. Stern, A. M. (2005). *Eugenic nation.* Berkeley, CA: University of California Press.

74. Broberg, G., & Roll-Hansen, N. (1996). *Eugenics and the welfare state: Sterilization policy in Denmark, Sweden, Norway, and Finland.* East Lansing, MI: Michigan State University Press.

    Robertson, J. (2001). Japan's first cyborg? Miss Nippon, eugenics and wartime technologies of beauty, body, and blood. *Body and Society, 7,* 1–34.

    Stepan, N. L. (1991). *The hour of eugenics: Race, gender, and nation in Latin America.* Ithaca, NY: Cornell University Press.

75. Kevles, D. J. (1995). *In the name of eugenics.* Cambridge, MA: Harvard University Press.

76. Allen, G. (1975). Genetics, eugenics and class struggle. *Genetics, 79*(supplement), 29–45; Hahn, N. (1980). Too dumb to know better: Cacogenic family studies and the criminology of women. *Criminology,*

*18*, 3–25; Rafter, N. (2008). *The criminal brain*. New York, NY: New York University Press; Stern, A. M. (2005). *Eugenic nation*. Berkeley, CA: University of California Press.

77. Selden, S. (2005). Transforming better babies into fitter families: Archival resources and the history of the American Eugenics Movement, 1908–1930. *Proceedings of the American Philosophical Society, 149*, 199–225.

78. Paul, D. B. (1995). *Controlling human heredity: 1865 to the present*. New York, NY: Humanity Books.

79. Singleton, M. M. (2014). The "science" of eugenics: America's moral detour. *Journal of American Physicians and Surgeons 19*(4), 122–125.

80. Dorr, G., & Logan, A. (2011). Quality, not mere quantity counts: Black eugenics and the NAACP baby contests. In P. Lombardo (Ed.), *A century of eugenics in America* (pp. 68–92). Bloomington, IN: Indiana University Press.

81. Shevell, M. (2012). A Canadian paradox: Tommy Douglas and eugenics. *Canadian Journal of Neurological Sciences, 39*, 35–39.

82. Spektorowski, A., & Ireni-Saban, L. (2013). *Politics of eugenics: Productionism, population, and national welfare*. New York, NY: Routledge.

    Paul, D. B. (1995). *Controlling human heredity: 1865 to the present*. New York, NY: Humanity Books.

83. Comfort, N. (2012). *The science of human perfection*. New Haven, CT: Yale University Press.

84. Wertz, D. (1998). Eugenics is alive and well: A survey of genetics professionals around the world. *Science in Context, 11*, 493–510.

85. Stern, A. M. (2005). *Eugenic nation: Faults and frontiers of better breeding in modern America*. Berkeley, CA: University of California Press.

    Also see Cockburn, A. (1997, October 2). A big green bomb aimed at immigration. *Los Angeles Times*. Retrieved from http://articles.latimes.com/1997/oct/02/local/me-38318.

86. Stern, A. M. (2005). *Eugenic nation: Faults and frontiers of better breeding in modern America*. Berkeley, CA: University of California Press; Paul, D. B. (1995). *Controlling human heredity: 1865 to the present*. New York, NY: Humanity Books.

    Kevles, D. J. (1995). *In the name of eugenics*. Cambridge, MA: Harvard University Press.

87. Ludmerer, K. M. (1972). *Genetics and American society: A historical appraisal* (p. 34). Baltimore, MD: Johns Hopkins University Press.

88. Comfort, N. (2012). *The science of human perfection*. New Haven, CT: Yale University Press.

89. Paul, D. B. (1995). *Controlling human heredity: 1865 to the present.* New York, NY: Humanity Books.

90. Paul, D. B. (1995). *Controlling human heredity: 1865 to the present.* New York, NY: Humanity Books.

91. Black, E. (2003). *War against the weak: Eugenics and America's campaign to create a master race.* New York, NY: Four Walls Eight Windows.

92. Black, E. (2003). *War against the weak: Eugenics and America's campaign to create a master race.* New York, NY: Four Walls Eight Windows; Paul, D. B. (1995). *Controlling human heredity: 1865 to the present.* New York, NY: Humanity Books; Kevles, D. J. (1995). *In the name of eugenics.* Cambridge, MA: Harvard University Press.

93. Wolstenholme, G. (1963). *Man and his future* (p. 275). Boston, MA: Little, Brown.

94. Bhattacharya, S. (2003, February 28). Stupidity should be cured, says DNA discoverer. *Daily News.* Retrieved from https://www.newscientist.com/article/dn3451-stupidity-should-be-cured-says-dna-discoverer/.

95. Comfort, N. (2012). *The science of human perfection.* New Haven, CT: Yale University Press.

96. Kevles, D. J. (1995). *In the name of eugenics* (p. 53). Cambridge, MA: Harvard University Press.

97. Goddard, H. H. (1920). *Feeble-mindedness: Its causes and consequences* (p. 437). New York, NY: MacMillan.

98. Walter, H. E. (1913). *Genetics: An introduction to the study of heredity* (p. 255). New York, NY: MacMillan.

99. Keller, J. (2005). In genes we trust: The biological component of psychological essentialism and its relationship to mechanisms of motivated social cognition. *Journal of Personality and Social Psychology, 88,* 686–702.

100. The correlations across different studies and different measures ranged between $r = .30$ and $r = -.50$. See Heine, S. J., Cheung, B. Y., & Ream, C. (2015, February 28). *Genetic attributions underlie people's attitudes towards criminal responsibility and eugenics.* Symposium presentation at the Society for Personality and Social Psychology, Long Beach, CA.

101. Duster, T. (2003). *Backdoor to eugenics.* New York, NY: Routledge.

## 8. A Brave New World: Engineering Better Essences

1. Chang, E. (August 5, 2009). *In China, DNA tests on kids ID genetic gifts, careers.* CNN. Retrieved from http://edition.cnn.com/2009/WORLD/asiapcf/08/03/china.dna.children.ability/.

2. Webborn, N., Williams, A., McNamee, M., Bouchard, C., Pitsiladis, Y., Ahmetov, I., et al. (2015). Direct-to-consumer genetic testing for predicting sports performance and talent identification: Consensus statement. *British Journal of Sports Medicine, 49*, 1486–1491.

3. Macur, J. (2008, November 29). Born to run? Little ones get test for sports gene. *New York Times*. Retrieved from http://www.nytimes .com/2008/11/30/sports/30genetics.html?_r=4&.

4. As one example, American International Biotechnology Services received such a letter for its testing of sports genes. Retrieved from http://www.fda.gov/MedicalDevices/ProductsandMedicalProcedures/ InVitroDiagnostics/ucm255348.htm.

5. Yang, N., MacArthur, D. G., Gulbin, J. P., Hahn, A. G., Beggs, A. H., Easteal, S., et al. (2003). ACTN3 genotype is associated with human elite athletic performance. *American Journal of Human Genetics, 73*, 627–631.

6. See the discussion by one of the researchers to first identify the link between ACTN3 and athletic performance. MacArthur, D. (2008, November 30). The ACTN3 sports gene test: What can it really tell you? *Wired Magazine*. Retrieved from http://www.wired.com/2008/11/ the-actn3-sports-gene-test-what-can-it-really-tell-you/.

7. See Webborn, N., Williams, A., McNamee, M., Bouchard, C., Pitsiladis, Y., Ahmetov, I., et al. (2015). Direct-to-consumer genetic testing for predicting sports performance and talent identification: Consensus statement. *British Journal of Sports Medicine, 49*, 1486–1491.

8. Schell, J., & Van Montagu, M. (1977). The Ti-plasmid of *Agrobacterium tumefaciens*, a natural vector for the introduction of NIF genes in plants? *Basic Life Sciences, 9*, 159–179.

9. Sanford, J. C., Klein, T. M., Wolf, E. D., & Allen, N. (1987). Delivery of substances into cells and tissues using a particle bombardment process. *Particulate Science and Technology, 5*, 27–37.

10. Gonsalves, D. (2004). Transgenic papaya in Hawaii and beyond. *AgBioForum, 7*(1&2), 36–40.

11. Ronald, P. C., & McWilliams, J. E. (2010, May 14). Genetically engineered distortions. *New York Times*. Retrieved from http://www.nytimes .com/2010/05/15/opinion/15ronald.html.

12. Hightower, R., Baden, C., Penzes, E., Lund, P., & Dunsmuir, P. (1991). Expression of antifreeze proteins in transgenic plants. *Plant Molecular Biology, 17*, 1013–1021.

13. Fedoroff, N. (2004). *Mendel in the kitchen*. Washington, DC: Joseph Henry Press.

14. Cyranoski, D. (2015, September 29). Gene-edited "micropigs" to be sold

as pets at Chinese institute. *Nature News*. Retrieved from http://www
.nature.com/news/gene-edited-micropigs-to-be-sold-as-pets-at-chinese
-institute-1.18448.

15. Ye, X., Al-Babili, S., Klöti, A., Zhang, J., Lucca, P., Beyer, P., et al. (2000).
Engineering the provitamin A (beta-carotene) biosynthetic pathway into
(carotenoid-free) rice endosperm. *Science*, *287*, 303–305.

16. Harmon, A. (2013, August 24). Golden rice: Lifesaver? *New York Times*.
Retrieved from http://www.nytimes.com/2013/08/25/sunday-review/
golden-rice-lifesaver.html?hp&_r=0.

17. Lemaux, P. G. (2008). Genetically engineered plants and foods: A scientist's analysis of the issues (Part I). *Annual Review of Plant Biology*,
*59*, 771–812.

18. Hallman, W., Hebden, C., Aquino, H., Cuite, C., & Lang, J. (2003).
*Public perceptions of genetically modified foods: A national study of
American knowledge and opinion*. Publication number RR-1003-004.
New Brunswick, NJ: Food Policy Institute, Cook College, Rutgers.

19. Harmon, A. (2014, January 4). A lonely quest for facts on genetically
modified crops. *New York Times*. Retrieved from http://www.nytimes
.com/2014/01/05/us/on-hawaii-a-lonely-quest-for-facts-about-gmos
.html?hp&_r=1.

20. Frewer, L. J., van der Lans, I. A., Fischer, A. R., Reinders, M. J., Menozzi,
D., Zhang, X., et al. (2013). Public perceptions of agri-food applications
of genetic modification—A systematic review and meta-analysis. *Trends
in Food Science & Technology*, *30*, 142–152.

21. Moon, W., & Balasubramanian, S. K. (2001). Public perceptions and
willingness-to-pay a premium for non-GM foods in the US and UK.
*AgBioForum*, *4*(3&4), 221–231.

22. The correlations between GMO support and the number of genetic
courses taken, as well as general knowledge of genetics, across a variety
of different samples ranged from .12 – .31. Ream, C., Cheung, B. Y., &
Heine, S. J. (2015). *Correlations with GMO attitudes*. Unpublished data.
University of British Columbia.

23. Hallman, W., Hebden, C., Aquino, H., Cuite, C., & Lang, J. (2003).
*Public perceptions of genetically modified foods: A national study of
American knowledge and opinion*. Publication number RR-1003-004.
New Brunswick, NJ: Food Policy Institute, Cook College, Rutgers.

24. Gaskell, G., Allum, N. C., & Stares, S. R. (2003). *Europeans and biotechnology in 2002: Eurobarometer 58.0*. Brussels: European Commission.

25. Hallman, W. K., Cuite, C. L., & Morin, X. K. (2013). *Public perceptions of labeling genetically modified foods* (Working paper 2013-01).

Retrieved from http://humeco.rutgers.edu/documents_pdf/news/gmlabelingperceptions.pdf.

26. Pew Research Center. (2015, January 29). *Chapter 3: Attitudes and beliefs on science and technology topics.* Retrieved from http://www.pewinternet.org/2015/01/29/chapter-3-attitudes-and-beliefs-on-science-and-technology-topics/.

27. American Association for the Advancement of Science (2012, October 20). Labeling of genetically modified foods. Retrieved from http://archives.aaas.org/docs/resolutions.php?doc_id=464.

28. For some reviews on the early artificial selection of wheat, see Anderson, P. C. (1999). *Prehistory of agriculture: New experimental and ethnographic approaches,* Los Angeles: University of California Press; Fedoroff, N. (2004). *Mendel in the kitchen.* Washington, DC: Joseph Henry Press.

29. Ahloowalia, B. S., Maluszynski, M., & Nichterlein, K. (2004). Global impact of mutation-derived varieties. *Euphytica, 135,* 187–204.

30. For a review, see Miller, H. I., & Conko, G. (2004). *The Frankenfood myth.* Westport, CT: Praeger.

31. Guyton, K. Z., Loomis, D., Grosse, Y., El Ghissassi, F., Benbrahim-Tallaa, L., Guha, N., et al. (2015). Carcinogenicity of tetrachlorvinphos, parathion, malathion, diazinon, and glyphosate. *The Lancet Oncology, 16,* 490–491.

32. Scott, S. E., Inbar, Y., & Rozin, P. (in press). Evidence for absolute moral opposition to genetically modified food in the United States. *Perspectives in Psychological Science.*

33. Baron, J., & Spranca, M. (1997). Protected values. *Organizational Behavior and Human Decision Processes, 70,* 1–16; Tetlock, P. E. (2003). Thinking the unthinkable: Sacred values and taboo cognition. *Trends in Cognitive Sciences, 7,* 320–324.

34. Haidt, J., & Hersh, M. A. (2001). Sexual morality: The cultures and emotions of conservatives and liberals. *Journal of Applied Social Psychology, 31,* 191–221.

    Haidt, J., Bjorklund, F., & Murphy, S. (1999). *Moral dumbfounding: When intuition finds no reason.* Unpublished manuscript, University of Virginia, Charlottesville, VA.

35. Harmon, A. (2013, August 24). Golden rice: Lifesaver? *New York Times.* Retrieved from http://www.nytimes.com/2013/08/25/sunday-review/golden-rice-lifesaver.html?hp&_r=0.

36. Scott, S. E., Inbar, Y., & Rozin, P. (in press). Evidence for absolute moral opposition to genetically modified food in the United States. *Perspectives in Psychological Science.*

37. Rozin, P., Spranca, M., Krieger, Z., Neuhaus, R., Surillo, D., Swerdlin, A., & Wood, K. (2005). Preference for natural: Instrumental and ideational/moral motivations, and the contrast between foods and medicines. *Appetite, 43*, 147–154.

38. Tenbült, P., de Vries, N. K., Dreezens, E., & Martijn, C. (2005). Perceived naturalness and acceptance of genetically modified food. *Appetite, 45*, 47–50.

39. See Condit, C. M. (2010). Public attitudes and beliefs about genetics. *Annual Review Genomics and Human Genetics, 11*, 339–359.

40. Specter, M. (2014, August 25). Seeds of doubt. *The New Yorker.* Retrieved from http://www.newyorker.com/magazine/2014/08/25/seeds-of-doubt.

41. Pew Research Center. (2015, January 29). *Chapter 3: Attitudes and beliefs on science and technology topics.* Retrieved from http://www.pewinternet.org/2015/01/29/chapter-3-attitudes-and-beliefs-on-science-and-technology-topics/.

42. Harris Poll (2014). Doctors, military officers, firefighters, and scientists seen as among America's most prestigious occupations. Retrieved from http://www.theharrispoll.com/politics/Doctors__Military_Officers__Firefighters__and_Scientists_Seen_as_Among_America_s_Most_Prestigious_Occupations.html.

43. Gaskell, G., Stares, S., Allansdottir, A., Allum, N., Castro, P., Esmer, Y., et al. (2010). *Europeans and biotechnology in 2010: Winds of change?* (Eurobarometer Survey Series No. 7; Report EUR 24537). Luxembourg: Publications Office of the European Union. Retrieved from http://ec.europa.eu/public_opinion/archives/ebs/ebs_341_winds_en.pdf.

44. Hallman, W., Hebden, C., Cuite, C., Aquino, H., & Lang, J. (2004). *Americans and GM food: Knowledge, opinion and interest in 2004.* Publication number RR-1104-007. New Brunswick, NJ: Food Policy Institute, Cook College, Rutgers.

45. For a cogent critique of the arguments by some GMO critics, see Lemaux, P. G. (2008). Genetically engineered plants and foods: A scientist's analysis of the issues (Part I). *Annual Review of Plant Biology, 59*, 771–812. Also see Saletan, W. (2015, July 15). Unhealthy fixation. The war against genetically modified organisms is full of fearmongering, errors, and fraud. Labeling them will not make you safer. *Slate.* Retrieved from http://www.slate.com/articles/health_and_science/science/2015/07/are_gmos_safe_yes_the_case_against_them_is_full_of_fraud_lies_and_errors.html.

46. For an egregious example of several of these nonexistent studies purporting dangers, see Smith, J. M. (2007). *Genetic roulette: The documented health risks of genetically engineered foods.* Fairfield, IA: Yes! Books.

47. Séralini, G.-E., Clair, E., Mesnage, R., Gress, S., Defarge, N., Malatesta, M., et al. (2012). Long-term toxicity of a Roundup herbicide and a Roundup-tolerant genetically modified maize. *Food and Chemical Toxicology*, *50*, 4221–4231.

48. Hayes, A. W. (2014). Retraction notice to "Long-term toxicity of a Roundup herbicide and a Roundup-tolerant genetically modified maize." *Food and Chemical Toxicology*, *63*, 244.

49. Wilson, R. D., Langlois, S., & Johnson, J. (2007). Mid-trimester amniocentesis fetal loss rate. *Journal of Obstetrics and Gynaecology Canada*, *194*, 586–590.

50. Sayres, L. C., & Cho, M. K. (2011). Cell-free fetal nucleic acid testing: A review of the technology and its applications. *Obstetrical & Gynecological Survey*, *66*, 431–442.

   Norton, M. E., & Wapner, R. J. (2015). Cell-free DNA analysis for noninvasive examination of trisomy. *The New England Journal of Medicine*, *372*, 1589–1597.

51. Mansfield, C., Hopfer, S., & Marteau, T. M. (1999). Termination rates after prenatal diagnosis of Down syndrome, spinal bifida, anencephaly, and Turner and Klinefelter syndromes: A systematic literature review. *Prenatal Diagnosis*, *19*, 808–812.

52. Natoli, J. L., Ackerman, D. L., McDermott, S., & Edwards, J. G. (2012). Prenatal diagnosis of Down syndrome: A systematic review of termination rates (1995–2011). *Prenatal Diagnosis*, *32*, 142–153.

53. Retrieved from http://www.omim.org.

54. Retrieved from https://www.jewishgenetics.org/dor-yeshorim.

55. For an excellent discussion on Dor Yeshorim, see Gessen, M. (2008). *Blood matters*. New York, NY: Mariner Books.

56. Appel, J. M. (2009, March 4). Mandatory genetic testing isn't eugenics, it's smart science. *Opposing Views*. Retrieved from http://www.opposingviews.com/i/mandatory-genetic-testing-isn-t-eugenics-it-s-smart-science.

57. Savulescu, J. (2001). Procreative beneficence: Why we should select the best children. *Bioethics*, *15*, 413–426 (p. 415).

58. Hudson, K., & Scott, J. (2002). *Public awareness and attitudes about genetic technology*. Washington, DC: Genetics and Public Policy Center.

59. Wasserman, D., & Asch, A. (2006). The uncertain rationale for prenatal disability screening. *AMA Journal of Ethics*, *8*, 1, 53–56.

60. Quoted in Abraham, C. (2012, January 7). Unnatural selection: Is evolving reproductive technology ushering in a new age of eugenics? *The Globe and Mail*. Retrieved from http://www.theglobeandmail.com/

life/parenting/unnatural-selection-is-evolving-reproductive-technology-ushering-in-a-new-age-of-eugenics/article1357885/?page=all.

61. Winkelman, W. D., Missmer, S. A., Myers, D., & Ginsburg, E. S. (2015). Public perspectives on the use of preimplantation genetic diagnosis. *Journal of Assisted Reproductive Genetics, 32,* 665–675.

62. Hudson, K., & Scott, J. (2002). *Public awareness and attitudes about genetic technology.* Washington, DC: Genetics and Public Policy Center.

63. For a riveting story about Molly Nash, see Faison, A. M. (2005, August). The miracle of Molly. *5280: The Denver Magazine.* Retrieved from http://www.5280.com/magazine/2005/08/miracle-molly?page=full.

64. Rivard, L. (2013, June 11). Case study in savior siblings. *Scitable by Nature Education.* Retrieved from http://www.nature.com/scitable/forums/genetics-generation/case-study-in-savior-siblings-104229158.

65. Spriggs, M. (2002). Lesbian couple create a child who is deaf like them. *Journal of Medical Ethics, 28,* 283.

66. Baruch, S. (2008). Preimplantation genetic diagnosis and parental preferences: Beyond deadly disease. *Houston Journal of Health Law & Policy, 8,* 245–270.

67. Baruch, S., Kaufman, D., & Hudson, K. L. (2008). Genetic testing of embryos: Practices and perspectives of US in vitro fertilization clinics. *Fertility & Sterility, 89,* 1053–1058.

68. Winkelman, W. D., Missmer, S. A., Myers, D., & Ginsburg, E. S. (2015). Public perspectives on the use of preimplantation genetic diagnosis. *Journal of Assisted Reproductive Genetics, 32,* 665–675.

69. Savulescu, J. (2001). Procreative beneficence: Why we should select the best children. *Bioethics, 15,* 413–426.

70. Byrne, D. (1971). *The attraction paradigm.* New York: Academic Press.

71. Berscheid, E., & Dion, K. (1971). Physical attractiveness and dating choice: A test of the matching hypothesis. *Journal of Experimental Social Psychology, 7,* 173–189.

    Kalick, S. M., & Hamilton, T. E. III (1986). The matching hypothesis reexamined. *Journal of Personality and Social Psychology, 51,* 673–682.

72. Spar, D. L. (2006). *The baby business.* Boston, MA: Harvard Business School Press.

73. Retrieved from http://www.cryobank.com.

74. Quoted on p. 7 of Plotz, D. (2006). *The genius factory: The curious history of the Nobel prize sperm bank.* New York, NY: Random House.

75. Plotz, D. (2006). *The genius factory: The curious history of the Nobel prize sperm bank.* New York, NY: Random House.

76. Scheib, J. E. (1994). Sperm donor selection and the psychology of female mate choice. *Ethology & Sociobiology* (now *Evolution and Human Behavior*), *15*, 113–129.

    Scheib, J. E., Kristiansen, A., & Wara, A. (1997). A Norwegian note on sperm donor selection and the psychology of female mate choice. *Evolution and Human Behavior, 18*, 143–149.

77. The Ethics Committee of the American Society for Reproductive Medicine (2007). Financial compensation of oocyte donors. *Fertility and Sterility, 88*(2), 305–309. Retrieved from https://www.asrm.org/uploadedFiles/ASRM_Content/News_and_Publications/Ethics_Committee_Reports_and_Statements/financial_incentives.pdf.

78. Spar, D. L. (2006). *The baby business.* Boston, MA: Harvard Business School Press.

79. Levine, A. D. (2010, March–April). Self-regulation, compensation, and the ethical recruitment of oocyte donors. *The Hastings Center Report, 40*(2), 25–36.

80. Levine, A. D. (2010, March–April). Self-regulation, compensation, and the ethical recruitment of oocyte donors. *The Hastings Center Report, 40*(2), 25–36.

81. For some examples of the hysteria, see Nichol, D. (1997, February 28). Hello Dolly, the Scottish designer sheep. *The Globe and Mail.* A. 19; Cohen, R. (1997, March 30). Hello, Dolly. *The Washington Post.* W04; Kolata, G. (1997, February 24). With cloning of a sheep, the ethical ground shifts. *New York Times.* Front page; Anonymous. (1997, March 1). Clone the clowns. *The Economist, 341,* 80. For more crazy examples, see p. 197 of Nelkin, D., & Lindee, M. S. (2004). *The DNA mystique.* Ann Arbor, MI: University of Michigan Press; and p. 145 of Condit, C. M. (1999). *The meanings of the gene.* Madison, WI: University of Wisconsin Press.

82. Condit, C. M. (2010). Public attitudes and beliefs about genetics. *Annual Review of Genomics and Human Genetics, 11,* 339–359; Calnan, M., Montaner, D., & Horne, R. (2005). How acceptable are innovative health-care technologies? A survey of public beliefs and attitudes in England and Wales. *Social Science & Medicine, 60,* 1937–1948; Shepherd, R., Barnett, J., Cooper, H., Coyle, A., Moran-Ellis, J., Senior, V., et al. (2007). Towards an understanding of British public attitudes concerning human cloning. *Social Science & Medicine, 65,* 377–392;

    The Wellcome Trust (1998). *Public perspectives on human cloning.* Medicine Society Programme (www.wellcome.ac.uk).

83. Tachibana, M., Amato, P., Sparman, M., Gutierrez, N. M., Tippner-Hedges, R., Ma, H., et al. (2013). Human embryonic stem cells derived by somatic cell nuclear transfer. *Cell, 153*, 1228–1238.

84. Shin, T., Kraemer, D., Pryor, J., Liu, L., Rugila, J., Howe, L., et al. (2002). Brief communications. Cell biology: A cat cloned by nuclear transplantation. *Nature, 415*, 859. Also see Francis, R. C. (2011). Epigenetics: The ultimate mystery of inheritance. New York, NY: W. W. Norton; and Yin, Sophia (2011, June 21). *Cloning cats: Rainbow and CC prove that cloning won't resurrect your pet* [Blog post]. Retrieved from http://drsophiayin.com/blog/entry/cloning-cats-rainbow-and-cc-prove-that-cloning-wont-resurrect-your-pet.

85. Ishino, Y., Shinagawa, H., Makino, K., Amemura, M., & Nakata, A. (1987). Nucleotide sequence of the iap gene, responsible for alkaline phosphatase isozyme conversion in *Escherichia coli*, and identification of the gene product. *Journal of Bacteriology, 169*, 5429–5433.

86. Barrangou, R., Fremaux, C., Deveau, H., Richards, M., Boyaval, P., Moineau, S., et al. (2007). CRISPR provides acquired resistance against viruses in prokaryotes. *Science, 315*, 1709–1712; Hsu, P. D., Lander, E. S., & Zhang, F. (2014). Development and applications of CRISPR-Cas9 for genome engineering. *Cell, 157*, 1262–1278; Jinek, M., Chylinski, K., Fonfara, I., Hauer, M., Doudna, J. A., & Charpentier, E. (2012). A programmable dual-RNA-guided DNA endonuclease in adaptive bacterial immunity. *Science, 337*, 816–821. For a couple of nice reviews of CRISPR in lay language, see Pollack, A. (2015, May 11). Jennifer Doudna, a pioneer who helped simplify genome editing. *New York Times.* Retrieved from http://www.nytimes.com/2015/05/12/science/jennifer-doudna-crispr-cas9-genetic-engineering.html?smid=pl-share; and Kahn, J. (2015, November 9). The Crispr quandary. *New York Times.* Retrieved from http://www.nytimes.com/2015/11/15/magazine/the-crispr-quandary.html?action=click&pgtype=Homepage&region=CColumn&module=MostEmailed&version=Full&src=me&WT.nav=MostEmailed.

87. Baltimore, D., Berg, P., Botchan, M., Carroll, D., Charo, R. A., Church, G., et al. (2015). A prudent path forward for genomic engineering and germline gene modification. *Science, 348*, 36–38.

88. Liang, P., Xu, Y., Zhang, X., Ding, C., Huang, R., Zhang, Z., et al. (2015). CRISPR/Cas9-mediated gene editing in human tripronuclear zygotes. *Protein & Cell, 6*(5), 363–372.

89. Cyranoski, D., & Reardon, S. (2015, April 22). Chinese scientists genetically modify human embryos. *Nature News.* Retrieved from http://www.nature.com/news/chinese-scientists-genetically-modify-human-embryos-1.17378

90. Cressey, D., Abbott, A., & Ledford, H. (2015, September 18). UK scientists apply for licence to edit genes in human embryos. *Nature News*. Retrieved from http://www.nature.com/news/uk-scientists -apply-for-licence-to-edit-genes-in-human-embryos-1.18394.

91. Yang, L., Güell, M., Niu, D., George, H., Lesha, E., Grishin, D., et al. (2015). Genome-wide inactivation of porcine endogenous retroviruses (PERVs). *Science, 350,* 1101–1104.

92. A conference was held in December 2015, with the goal of drafting ethical guidelines of CRISPR research. Retrieved from https:// innovativegenomics.org/international-summit-on-human-gene-editing/.

93. Quote comes from episode 5 of the PBS documentary *DNA: Pandora's Box*. Retrieved from https://www.youtube.com/watch?v=qe4EW3AOgzs

94. Metzl, J. F. (2014, October 10). The genetics epidemic: The revolution in DNA science and what to do about it. *Foreign Affairs*. Retrieved from https://www.foreignaffairs.com/articles/united-states/2014-10-10/ genetics-epidemic.

  Miller, G. (2013). Chinese eugenics. *The Edge*. Retrieved from http:// edge.org/response-detail/23838.

95. Winkelman, W. D., Missmer, S. A., Myers, D., & Ginsburg, E. S. (2015). Public perspectives on the use of preimplantation genetic diagnosis. *Journal of Assisted Reproductive Genetics, 32,* 665–675.

  Hudson, K., & Scott, J. (2002). *Public awareness and attitudes about genetic technology*. Washington, DC: Genetics and Public Policy Center.

96. Duster, T. (2003). *Backdoor to eugenics*. New York, NY: Routledge.

97. Schwartz, B., Ward, A., Monterosso, J., Lyubomirsky, S., White, K., & Lehman, D. R. (2002). Maximizing versus satisficing: Happiness is a matter of choice. *Journal of Personality and Social Psychology, 83,* 1178–1197.

98. Schwartz, B., Ward, A., Monterosso, J., Lyubomirsky, S., White, K., & Lehman, D. R. (2002). Maximizing versus satisficing: Happiness is a matter of choice. *Journal of Personality and Social Psychology, 83,* 1178–1197.

99. Power, R. A., Steinberg, S., Bjornsdottir, G., Rietveld, C. A., Abdellaoui, A., Nivard, M. M., et al. (2015). Polygenic risk scores for schizophrenia and bipolar disorder predict creativity. *Nature Neuroscience, 18,* 953–955; Andreasen, N. C. (1987). Creativity and mental illness: Prevalence rates in writers and their first-degree relatives. *American Journal of Psychiatry, 144,* 1288–1292.

## 9. Harnessing Our Essentialist Impulses: How Should We Think about Genes?

1. Brenner, S. (2009; October 30). The personalized genome: Do I want to know? Gairdner Foundation 50th Anniversary, Toronto, ON. Retrieved from http://mediacast.ic.utoronto.ca/20091030-GFS-P4/index.htm.

2. The items come from Forer, B. R. (1949). The fallacy of personal validation: A classroom demonstration of gullibility. *Journal of Abnormal and Social Psychology, 44*, 118–123.

3. Smoller, J. W., Paulus, M. P., Fagerness, J. A., Purcell, S., Yamaki, L. H., Hirshfeld-Becker, D., et al. (2008). Influence of RGS2 on anxiety-related temperament, personality, and brain function. *Archives of General Psychiatry, 65*, 298–308.

4. Retrieved from https://www.23andme.com/you/community/thread/6235/.

5. Pinker, S. (2009, January 7). My genome, my self. *New York Times.* Retrieved from http://www.nytimes.com/2009/01/11/magazine/11Genome-t.html?_r=0.

6. See page xxiii of Collins, F. S. (2010). *The language of life.* New York, NY: Harper Perennial.

7. Pinker, S. (2009, January 7). My genome, my self. *New York Times.* Retrieved from http://www.nytimes.com/2009/01/11/magazine/11Genome-t.html?_r=0.

8. http://www.improbable.com/hair/gallery2/.

9. Austin, J. C. (2010). Re-conceptualizing risk in genetic counseling: Implications for clinical practice. *Journal of Genetic Counseling, 19*, 228–234.

10. Evans, D. G. R., Blair, V., Greenhalgh, R., Hopwood, P., & Howell, A. (1994). The impact of genetic counseling on risk perception in women with a family history of breast cancer. *British Journal of Cancer, 70*, 934–938.

    Smerecnik, C. M. R., Mesters, I., Verweij, E., de Vries, N. K, & de Vries, H. (2009). A systematic review of the impact of genetic counseling on risk perception accuracy. *Journal of Genetic Counseling, 18*, 217–228.

11. Quote from Sheldon Reed reported in Stern, A. M. (2012). *Telling genes* (p. 47). Baltimore, MD: Johns Hopkins University Press.

12. Rothman, B. K. (1986). *The tentative pregnancy* (p. 43). New York, NY: Viking.

13. Quote from Diana Punales-Morejon reported in Stern, A. M. (2012). *Telling genes* (p. 40). Baltimore, MD: Johns Hopkins University Press.

14. Cheung, B. Y. (2016). *Out of my control: The effects of perceived genetic etiology.* Unpublished doctoral dissertation: University of British Columbia.

15. Faraone, S. V., Smoller, J. W., Pato, C. N., Sullivan, P., & Tsuang, M. T. (2008). Editorial: The new neuropsychiatric genetics. *American Journal of Medical Genetics Part B (Neuropsychiatric Genetics), 147B,* 1–2.

16. Benjamin, D. J., Cesarini, D., van der Loos, M. J. H. M, Dawes, C. T., Koellinger, P. D., Magnusson, P. K. E., et al. (2012). The genetic architecture of economic and political preferences. *Proceedings of the National Academy of Science, 109,* 8026–8031.

17. For a cutting critique of the inherent value of GWAS studies, see Goldstein, D. B. (2009). Common genetic variation and human traits. *New England Journal of Medicine, 360,* 1696–1698.

18. Dobbs, D. (2015, May 21). What is your DNA worth? *BuzzFeed.* Retrieved from http://www.buzzfeed.com/daviddobbs/weighing-the -promises-of-big-genomics#.vjnnjJzwK.

19. Wang, K., Zhang, H., Ma, D., Bucan, M., Glessner, J. T., Abrahams, B. S., et al. (2009). Common genetic variants on 5p14.1 associate with autism spectrum disorders. *Nature, 459,* 528–533.

20. Bloomberg News (2009, April 29). Researchers identify autism genes that give clues to brain structure. *New York Daily News.* Retrieved from http://www.nydailynews.com/life-style/health/researchers-identify -autism-genes-give-clues-brain-structure-article-1.360034.

21. Weiss, L. A. (2009). A genome-wide linkage and association scan reveals novel loci for autism. *Nature, 461,* 802–808.

22. For a list of the effects this gene has been linked to, see Charney, E., & English, W. (2012). Candidate genes and political behavior. *American Political Science Review, 106,* 1–34.

23. The effect size of the link with novelty-seeking is $d = .06$, which corresponds to an $R^2$ of slightly less than .001. See Kluger, A. N., Siegfried, Z., & Ebstein, R. P. (2002). A meta-analysis of the association between DRD4 polymorphism and novelty seeking. *Molecular Psychiatry, 7,* 712–717.

24. The reason why this is an overestimate is that there are not just two variants of the number of repeats of this variable region of the gene, which affect how it influences dopamine signaling, but there are at least 10 variants, ranging from 2 repeats to the much rarer 11 repeats. The problem is that different studies call different versions of these alleles as high or low in dopamine signaling, which confounds a comparison of them. Some studies only call the 7 repeat polymorphism as high in dopamine signaling in contrast to all other variants (e.g., Ebstein, R. P., et al. [1996]); some group 6, 7, and 8 repeats and con-

trast these to all others (e.g., Swift et al. [2000]); some group 2, 3, 4, and 5 together, in contrast to 6, 7, and 8 repeats; e.g., Vadenberg et al. [1997]); some group 5 and 6 together and contrast them with 2, 3, and 4 repeats (e.g., Ono et al. [1997]); some group 2, 3, 4, 5, and 6 and contrast them to 7 or more repeats (e.g., Garcia, J. R., et al. [2010]); some group 2, 3, and 4, in contrast to 5, 6, and 7 repeats (e.g., Tomitaka et al. [1999]); some contrast 2, 3, and 4 repeats with 5, 6, 7, 8, 9, 10, and 11 repeats (e.g., Chen et al., 1999); some contrast 7 and 2 repeats with all others (Reist et al. [2007]), some contrast 7 and 2 repeats with 4 repeats (Kitayama et al. [2014]). Chen, C., Burton, M., Greenberger, E., & Dmitrieva, J. (1999). Population migration and the variation of dopamine D4 receptor (DRD4) allele frequencies around the globe. *Evolution and Human Behavior, 20,* 309–324; Ebstein, R. P., et al. (1996). Dopamine D4 receptor (DRD4) exon III polymorphism associated with the human personality trait of novelty seeking. *Nature Genetics, 12,* 78–80; Garcia, J. R., MacKillop, J., Aller, E. L., Merriwether, A. M., Wilson, D. S., & Lum, J. K. (2010). Associations between dopamine D4 receptor gene variation with both infidelity and sexual promiscuity. *PLoS ONE 5(11):* e14162; Kitayama, S., King, A., Yoon, C., Tompson, S., Huff, S., & Liberzon, I. (2014). The dopamine D4 receptor gene (DRD4) moderates cultural difference in independent versus interdependent social orientation. *Psychological Science, 25,* 1169–1177; Ono, Y., Manki, H., Yoshimura, K., Muramatsu, T., Mizushima, H., Higuchi, S., et al. (1997). Association between dopamine D4 receptor (D4DR) exon III polymorphism and novelty seeking in Japanese subjects. *American Journal of Medical Genetics, 74,* 501–503; Reist, C., Ozdemir, V., Wang, E., Hashemzadeh, M., Mee, S., & Moyzis, R. (2007). Novelty seeking and the dopamine D4 receptor gene (DRD4) revisited in Asians. *American Journal of Medical Genetics Part B, 144B,* 453–457; Swift, G., Larsen, B., Hawi, Z., & Gill, J. (2000). Novelty seeking traits and D4 dopamine receptors. *American Journal of Medical Genetics, 96,* 222–223; Tomitaka, M., Tomitaka, S., Otuka, Y., Kim, K., Matuki, H., Sakamoto, K., et al. (1999). Association between novelty seeking and dopamine receptor D4 (DRD4) exon III polymorphism in Japanese subjects. *American Journal of Medical Genetics, 88,* 469–471; V Vandenbergh, D. J., Zonderman, A. B., Wang, J., Uhl, G. R., & Costa, P. T. Jr. (1997). No association between novelty seeking and dopamine D4 receptor (D4DR) exon III seven repeat alleles in Baltimore Longitudinal Study of Aging participants. *Molecular Psychiatry, 2,* 417–419.

25. For an elaboration of this point, see Murray, A. B., Carson, M. J., Morris,

C. A., & Beckwith, J. (2010). Illusions of scientific legitimacy: misrepresented science in the direct-to-consumer genetic-testing marketplace. *Trends in Genetics*, *26*(11), 459–461.

26. Autism Developmental Disabilities Monitoring Network Surveillance Year 2010. Principle Investigators (2014). Prevalence of autism spectrum disorder among children aged 8 years. *Morbidity and Mortality Weekly Report*, *63*(SS02), 1–21.

27. Berg, J. M., & Geschwind, D. H. (2012). Autism genetics: Searching for specificity and convergence. *Genome Biology*, *13*, 247–263.

28. Sanders, S. J., Ercan-Sencicek, A. G., Hus, V., Luo, R., Murtha, M. T., Moreno-De-Luca, D., et al. (2011). Multiple recurrent de novo CNVs, including duplications of the 7q11.23 Williams syndrome region, are strongly associated with autism. *Neuron*, *70*, 863–885; Levy, D., Ronemus, M., Yamrom, B., Lee, Y. H., Leotta, A., Kendall, J., et al. (2011). Rare de novo and transmitted copy-number variation in autistic spectrum disorders. *Neuron*, *70*, 886–897; Gilman, S. R., Iossifov, I., Levy, D., Ronemus, M., Wigler, M., & Vitkup, D. (2011). Rare de novo variants associated with autism implicate a large functional network of genes involved in formation and function of synapses. *Neuron*, *70*, 898–907; Anney, R., Klei, L., Pinto, D., Almeida, J., Bacchelli, E., Baird, G., et al. (2012). Individual common variants exert weak effects on the risk for autism spectrum disorders. *Human Molecular Genetics*, *21*, 4781–4792; Kong, A., Frigge, M. L., Masson, G., Besenbacher, S., Sulem, P., Magnusson, G., et al. (2012). Rate of de novo mutations and the importance of father's age to disease risk. *Nature*, *488*, 471–475; Chan, J. A. (2015). The emerging picture of autism spectrum disorder: Genetics and pathology. *Annual Review of Pathology: Mechanisms of Disease*, *10*, 111–144.

29. Weintraub, K. (2011). Autism counts. *Nature*, *479*, 22–24.

30. See Gruber, J. (2013). The unfulfilled promise of genomics. In S. Krimsky & G. Gruber (Eds.) *Genetic explanations* (pp. 270–282). Cambridge, MA: Harvard University Press.

31. Joseph, J., & Ratner, C. (2013). The fruitless search for genes in psychiatry and psychology. In S. Krimsky & G. Gruber (Eds.) *Genetic explanations* (pp. 96–97). Cambridge, MA: Harvard University Press. Also see Chaufan, C. (2007). How much can a large population study on genes, environments, and their interactions and common diseases contribute to the health of the American people? *Social Science & Medicine*, *65*, 1730–1741.

32. Comfort, N. (2014, January 29). Genetic determinism: Why we never learn—and why it matters [Blog post]. Retrieved from http://genotopia.scienceblog .com/387/genetic-determinism-why-we-never-learn-and-why-it-matters/.

33. Pearson, H. (2009). Human genetics: One gene, twenty years. *Nature*, *460*, 164–169.

34. Amaral, M. D. (2015). Novel personalized therapies for cystic fibrosis: Treating the basic defect in all patients. *Journal of Internal Medicine*, *277*, 155–166.

35. Quote comes from p. 165 in Pearson, H. (2009). Human genetics: One gene, twenty years. *Nature*, *460*, 164–169.

36. For some discussions of the slow pace of genome-based treatments, see Palmer, B. (2013, September 30). Where are all the miracle drugs? *Slate Magazine*. Retrieved from http://www.slate.com/articles/health_ and_science/human_genome/2013/09/human_genome_drugs_where_ are_the_miracle_cures_from_genomics_did_the_genome.1.html; and Wade, N. (2010, June 12). A decade later, genetic map yields few new cures. *New York Times*. Retrieved from http://www.nytimes .com/2010/06/13/health/research/13genome.html?_r=0.

37. Chapman, P. B., Hauschild, A., Robert, C., Haanen, J. B., Ascierto, P., Larkin, J., et al. (2011). Improved survival with vemurafenib in melanoma with BRAF V600E mutation. *New England Journal of Medicine*, *364*, 2507–2516.

38. O'Brien, S. G. (2003). Imatinib compared with interferon and low-dose cytarabine for newly diagnosed chronic-phase chronic myeloid leukemia. *New England Journal of Medicine*, *348*, 994–1004.

39. Rodin, J., & Langer, E. J. (1977). Long-term effects of a control-relevant intervention with the institutionalized aged. *Journal of Personality and Social Psychology*, *35*, 897–902.

40. Ornish, D., Brown, S. E., Scherwitz, L. W., Billings. J. H., Armstrong, W. T., Ports, T. A., et al. (1990). Can lifestyle changes reverse coronary heart disease? The Lifestyle Heart Trial. *The Lancet*, *336*, 129–133.

41. Diabetes Prevention Program Research Group (2002). Reduction in the incidence of Type 2 diabetes with lifestyle intervention or metformin. *New England Journal of Medicine*, *346*, 393–403.

42. Smyth, J. M., Stone, A. A., Hurewitz, A., & Kaell, A. (1999). Effects of writing about stressful experiences on symptom reduction in patients with asthma or rheumatoid arthritis. *Journal of the American Medical Association*, *281*, 1304–1309.

43. Creswell, J. D., Myers, H. F., Cole, S. W., & Irwin, M. R. (2009).

Mindfulness meditation training effects on CD4+ T lymphocytes in HIV-1 infected adults: A small randomized controlled trial. *Brain, Behavior, and Immunity, 23,* 184–188.

44. Spiegel, D., Bloom, J. R., Kraemer, H. C., & Gottheil, E. (1989). Effect of psychosocial treatment on survival of patients with metastatic breast cancer. *The Lancet, 334,* 888–891; Goodwin, P. J., Leszcz, M., Ennis, M., Koopmans, J., Vincent, L., Guther, H., et al. (2001). The effect of group psychosocial support on survival in metastatic breast cancer. *The New England Journal of Medicine, 345,* 1719–1726.

45. Kluger, J. (2010, December 2). Too many one-night stands? Blame your genes. *Time.* Retrieved from http://healthland.time.com/2010/12/02/too-many-one-night-stands-blame-your-genes/; Firth, N., & Macrae, F. (2010, December 3). The Love-Cheat Gene: One in four born to be unfaithful, claim scientists. *Daily Mail.* Retrieved from http://www.dailymail.co.uk/sciencetech/article-1334932/The-love-cheat-gene-One-born-unfaithful-claim-scientists.html.

46. A nice discussion of this "scientists find genes for" point can be seen in a lecture by Dr. Steve Jones on "Nature or Nurture." Retrieved from https://www.youtube.com/watch?v=1ksP34GYwbY.

47. Conrad, P. (2002). Genetics and behavior in the news: Dilemmas of a rising paradigm. In J. S. Alper, C. Ard, A. Asch, J. Beckwith, P. Conrad, & L. N. Geller (Eds.) *The double-edged helix: Social implications of genetics in a diverse society.* Baltimore, MD: Johns Hopkins University Press.

48. Lakoff, G., & Johnson, M. (1980). *Metaphors we live by.* Chicago, IL: University of Chicago Press; Landau, M. J., Meier, B. P., & Keefer, L. A. (2010). A metaphor-enriched social cognition. *Psychological Bulletin, 136,* 1045–1067.

49. For some discussion on existing gene metaphors, see Condit, C. M. (1999). *The meanings of the gene.* Madison, WI: University of Wisconsin Press; Francis, R. C. (2011). *Epigenetics.* New York, NY: W. W. Norton; paragraph 6 of Sapolsky, R. (1997, October). A gene for nothing. *Discover.* Retrieved from http://discovermagazine.com/1997/oct/agenefornothing1242; Hubbard, R., & Wald, E. (1997). *Exploding the gene myth: How genetic information is produced and manipulated by scientists, physicians, employers, insurance companies, educators and law enforcers.* Boston, MA: Beacon Press; Keller, E. F. (2000). *The century of the gene.* Cambridge, MA: Harvard University Press.

50. See recent research showing how learning about Mendelian accounts of genetics leads to more genetic determinism than learning about a

rich developmental account of how genes interact with their environments. Radick, G. (2016). Teach students the biology of their time. *Nature, 533,* 293.

51. See Walker, I., & Read, J. (2002). The differential effectiveness of psychosocial and biogenetic causal explanations in reducing negative attitudes toward "mental illness." *Psychiatry: Interpersonal and Biological Processes, 65,* 313–325; Boysen, G. A., & Gabreski, J. D. (2012). The effect of combined etiological information on attitudes about mental disorders associated with violent and nonviolent behaviors. *Journal of Social and Clinical Psychology, 31,* 852–877; Cheung, B. Y., & Heine, S. J. (2016). *Efforts to reduce essentialist responses to genetic accounts.* Unpublished data. The University of British Columbia.

52. Ream, C., Cheung, B. Y., & Heine, S. J. (2016). *The role of genetics education on genetic essentialism.* Unpublished data. The University of British Columbia; Castéra, J., & Clément, P. (2014). Teachers' conceptions about the genetic determinism of human behaviour: A survey in 23 countries. *Science & Education, 23,* 417–443.

53. For a great review on this, see Nisbett, R. E. (2015). *Mindware: Tools for smart thinking.* New York, NY: Farrar, Straus, and Giroux.

54. For example, see Devine, P. G., Forscher, P. S., Austin, A. J., & Cox, W. T. L. (2012). Long-term reduction in implicit race bias: A prejudice habit-breaking intervention. *Journal of Experimental Social Psychology, 48,* 1267–1278.

55. For a couple of examples, see Beaman, A. L., Barnes, P. J., Klentz, B., & McQuirk, B. (1978). Increasing helping rates through informational dissemination: Teaching pays. *Personality and Social Psychology Bulletin, 4,* 406–411; Johns, M., Schmader, T., & Martens, A. (2005). Knowing is half the battle: Teaching stereotype threat as a means of improving women's math performance. *Psychological Science, 16,* 175–179.

# Index

Page numbers followed by *n* refer to footnotes or endnotes.